Prochloron
A Microbial Enigma

Contributors

Randall S. Alberte—Department of Molecular Genetics and Cell Biology, The University of Chicago, Chicago, IL 60637, U.S.A.

Tineke Burger-Wiersma—Laboratory for Microbiology, University of Amsterdam, 1018 WS Amsterdam, The Netherlands

Lanna Cheng—Scripps Institution of Oceanography, University of California, San Diego, La Jolla, CA 92093, U.S.A.

Ralph A. Lewin—Scripps Institution of Oceanography, University of California, San Diego, La Jolla, CA 92093, U.S.A.

Hans C. P. Matthijs—Laboratory for Microbiology, University of Amsterdam, 1018 WS Amsterdam, The Netherlands

Luuc R. Mur—Laboratory for Microbiology, University of Amsterdam, 1018 WS Amsterdam, The Netherlands

Rosevelt L. Pardy—School of Life Sciences, University of Nebraska, Lincoln, NE 68588, U.S.A.

Erko Stackebrandt—Institut für allgemeine Mikrobiologie der Universität Kiel, 23 Kiel, West Germany

Hewson Swift—Department of Molecular Genetics and Cell Biology, The University of Chicago, Chicago, IL 60637, U.S.A.

F. Robert Whatley—Department of Plant Science, Oxford University, Oxford, OX1 3RA, England

Prochloron
A Microbial Enigma

Edited by

Ralph A. Lewin and Lana Cheng

Published with support from the
Phycological Society of America

CHAPMAN AND HALL
NEW YORK LONDON

First published in 1989 by
Chapman and Hall
an imprint of
Routledge, Chapman & Hall, Inc.
29 West 35 Street
New York, NY 10001

Published in Great Britain by

Chapman and Hall
11 New Fetter Lane
London EC4P 4EE

Printed in the United States of America

Library of Congress Cataloging in Publication Data

Prochloron, a microbial enigma / edited by Ralph A. Lewin and Lanna Cheng.
p. cm.
Bibliography: p.
Includes index.
ISBN 0-412-01901-9
1. Prochloron. I. Lewin, Ralph A. II. Cheng, Lanna.
QK569.P73P76 1989 89-7401
589.4—dc20 CIP

British Library Cataloguing in Publication Data

Prochloron.
1. Prochloron
I. Lewin, Ralph A. II. Cheng, Lanna
589.4′6

ISBN 0-412-01901-9

to our collaborators, who did most of the work

Contents

Abbreviations and Acronyms

ADPG	adenosine diphosphate glucose
alb.	albert (micro-Einstein of PAR per meter^2 per second)
APC	allophycocyanin
ATP	adenosine triphosphate
C-PE	C-phycoerythrin
CZAR	daily symbiont-fixed carbon used to support host respiration
DAB	diaminobenzidine
DAPI	4′,6-diamidino-2-phenyl-indole
DGG	digalactosyl diacylglycerol
DNA	deoxyribonucleic acid
EDTA	ethylene diamine tetraacetic acid
EF	exoplasmic face (of thylakoid)
G-C	guanine plus cytocine
G:C	ratio of guanine to cytosine
Gly-DH	glycolate dehydrogenase
GOGAT	glutamine oxoglutarate aminotransferase
GS	glutamine synthetase
HPLC	high-performance liquid chromatography
I_k	photosynthetic light-saturation level
I_c	photosynthetic light-compensation level
kD	kilodaltons
K_m	Michaelis-Menten constant
LHC	light-harvesting complex
LHC II	light-harvesting chlorophyll a/b complex
MGG	monogalactosyl diacyl glycerol
mRNA	mitochondrial ribonucleic acid
MW	molecular weight

NADP	nicotinamide adenine dinucleotide
NiR	nitrite reductase
NMR	nuclear magnetic resonance
NR	nitrate reductase
OAA	oxaloacetic acid
PAGE	polyacrylamide gel electrophoresis
PAR	photosynthetically active radiation
PAS	periodic acid–Schiff reaction
PC	phycocyanin
PE	phycoerythrin
PEP	phosphoenolpyruvate
PF	protoplasmic face (of thylakoid)
PG	phosphatidyl glycerol
PGA	phosphoglyceric acid
P-I	photosynthesis-irradiance relationship
P_{max}	minimum quantum flux that saturates photosynthesis
P:R	photosynthesis/respiration ratio
PSI	photosystem I
PSII	photosystem II
PSU	photosynthetic unit
PVPP	polyvinyl polypyrrolidone
Q_{10}	respiratory quotient
RNA	ribonucleic acid
RNase	ribonuclease
rRNA	ribosomal RNA
RUBISCO	ribulose 1,5-bisphosphate carboxylase-oxygenase
RuBP	ribulose bisphosphate
SDS	sodium dodecyl sulfate
SL	sulpholipid=sulphoquinovosyl diacylglycerol
TCA	trichloroacetic acid
TEM	transmission electron microscopy
UDPG	uridine diphosphate glucose
UV	ultraviolet light

Acknowledgments

We gratefully acknowledge financial support from the National Aeronautics and Space Administration (N.A.S.A.), the National Science Foundation, the National Geographic Society and the American Philosophical Society (U.S.A.), the Royal Society and the British Council (U.K.), the Foundation for Ocean Research, San Diego, and the Academic Senate, University of California, San Diego (U.S.A.) which helped to defray costs of research on the expeditions (see Appendix 1). We are also grateful to N.A.S.A. for funding the First International Prochloron Conference, held at La Jolla, California, in January 1983, which enabled most of us to meet, some for the first time, to exchange research information and plan new ventures (see Appendix 2). These participants, and many others who collaborated in various ways, made important contributions to our knowledge of *Prochloron*, as this volume abundantly testifies.

A grant from the Phycological Society of America helped to defray production costs of the color plates in this volume, and Sue Stultz did Trojan work on the word processor.

Chapter I

Introduction

Ralph A. Lewin and Lanna Cheng

In physics, the discovery of new (more properly, hitherto undetected) particles has often resulted from a search: like the discovery of America, their existence had been postulated but their actual existence awaited confirmation. In biology, new discoveries are rarely made in this way. The existence of an alga like *Prochloron*, as a putative ancestor of chloroplasts, had been postulated, but in fact its discovery was a consequence of fortuitous events. Green algal symbionts in didemnid ascidians had been known for decades to a few marine zoologists who had worked in coral reef areas, but nobody had bothered much about them. When we happened to find them, under boulders on a seashore in Baja California, Mexico, where we were taking part in a student expedition, we didn't bother much either at first, though they worried us a little. With our portable microscope we could see no nuclei in the cells, which, according to the dogma accepted at the time, indicated that they were blue-green algae—yet they didn't look blue-green. They were leaf-green, like green algae and higher plants. We made desultory attempts to grow them in culture, in variously enriched seawater media, but failed. (This proved to be a frustrating experience, all too frequently repeated on subsequent expeditions.) We collected enough for electron microscopy, though, and transmission electron microscopy (TEM) studies indicated that the cells were unequivocally prokaryotic. But we couldn't collect enough for even a preliminary examination of the chlorophylls and were not absolutely sure of the absence of the blue bilin pigments that normally characterize cyanophytes. We looked in vain for such algae nearer home, seeking a source that would provide fresh material for study in the laboratories of the Scripps Institution of Oceanography. And so we shelved the problem: after all, at that time we were mainly studying, respectively, the genetics of the green flagellate *Chlamydomonas* and the biology of the marine insect *Halobates*.

A couple of years later another opportunity arose to visit Baja California, when one of us contracted to study the irritating problem of biting midges on Isla San José, in the Gulf of California. We managed to collect enough of the paradoxical algal cells, from didemnids growing on the roots of mangroves, to permit us to examine their lipid pigments by paper chromatography. The separation of two green spots provided indications of chlorophylls *a* and *b*. (Innumerable pink spots, on our own skins, bore evidence of the activities and appetites of the local midges, but on this as on later expeditions we bore such inconveniences with fortitude.) A nagging possibility that some foreign chlorophyte had contaminated our field collections with the unexpected chlorophyll *b* could not be completely excluded, but this became increasingly improbable as, year by year on later expeditions, our harvests increased from microgram to milligram quantities and the evidence for two chlorophylls grew more convincing. We needed even more and purer preparations, however, if we were to be really sure.

A chance conversation during a visit to Hawaii put us on to the existence of another and better source of these algae, which in this case live inside the animals (*Diplosoma virens*) in virtual purity of species. A few hundred milligrams of the animals were scraped from concrete walls and dead coral, and with our purer preparations we confirmed our earlier provisional results. At the time of our first brief publication on the subject we had described the alga as a cyanophyte, but the protests of colleagues and the increased confidence that we had gained from our newer and more convincing pigment analyses finally encouraged us to assign these microbes to a new subclass of algae. This seemed a practical, indeed necessary, expedient, if only because it allowed us to make bolder applications for financial support to study *Prochloron* in greater detail.

We had known from the zoological literature that such symbiotic algae had been reported only from tropical coasts. Another expedition, primarily to study intertidal insects on the coastal reefs of Eniwetok Atoll in the Marshall Islands, enabled us to collect more material. During a sabbatical leave in Australia, where we sought antibiotics in marine algal breis, we were supplied frozen material of the giant didemnid *Lissoclinum patella*, which for the first time permitted us to collect many grams of *Prochloron* in a relatively pure state for later investigations elsewhere.

On a visit to Singapore, we were told of the presence of a variety of different green didemnids on offshore islets, and on the shore of one of those furthest out, Pulau Salu, we found quantities of *Lissoclinum patella*, containing more *Prochloron* cells than we could use. Alas, dredging or other perturbations of the sea in those parts have depleted the stocks of this wonderful animal (we didn't do so ourselves, of course), and on a more recent visit we could find none at all. But on an expedition to the Caroline Islands on the R/V "Alpha Helix," partly to study the ecology of *Halobates* and partly to continue our surveys for *Prochloron* symbionts, we found even more green didemnids, especially around Palau, and we then learned of the value of having a lyophilizer aboard. The preparation of freeze-dried samples, for eventual biochemical and even molecular biological studies in well-equipped laboratories at home, was thereby made possible. (We recommend that every marine research laboratory in the tropics have freeze-drying equipment.) Later, we shipped a lyophilizer to Palau and were thus in a position to collect many batches of *Prochloron* from different sources, to prepare and preserve them in different ways, and to make them available to specialists all over the world.

This, then, is how the field of prochlorophycology grew up and how it has been able to make so much progress in one short decade. We still cannot grow *Prochloron,* but we now know a lot about its ecology, its physiology *in hospite,* and its biochemistry *in vitro,* to supplement all kinds of field observations that we and our colleagues have been able to carry out on our various expeditions. There have been 10 "expeditions," so far; a list of them, with the places visited and the names of the participants, is appended. The results of collaborative studies made in this way, and publications from a score of laboratories scattered on four continents, are summarized in the following pages. We have tried to include abstracts of all the papers dealing specifically with *Prochloron.* (Publications relating to *Prochloron* peripherally, but dealing mainly with its animal hosts, have been included in another list.) They were mostly published between 1976 and 1988, when the theory of symbiogenesis—postulating that green plant chloroplasts had developed from prokaryotic green symbionts—was regaining acceptance and respectability in the textbooks. Some of the funding for our research came from organizations specifically interested in testing this hypothesis. At the end of this decade, some of us are still far from convinced, but we think we have put some of the arguments, pro or con, on sounder factual bases. Our present inclination is to vote in the "con" lobby, because we feel that almost all the objective evidence indicates closer affinities between *Prochloron* and certain contemporary cyanophytes (Lewin, 1986). However, there is an Italian saying, *"Si non é vero, é ben' trovato,"* meaning that some stories, even if untrue, may nevertheless have the ring of plausibility. Even if ancestral prochlorophytes were not the direct forebears of green-plant chloroplasts, they surely had certain features in common. For this reason, if for no other, *Prochloron* merits further study. Possible origins of *Pro-*

chloron, and its involvement in the origins of other kinds of green algae, are presented diagrammatically in Figs. 1.1 through 1.6 (see Lewin, 1987, where various implications are discussed). Its resemblance to certain cyanophytes, found in similar coral-reef habitats, is certainly worth further consideration. (See Kott, 1984; Monniot, 1984; Olson, 1986; Parry, 1984; Ruetzler, 1981; Sybesma et al., 1981).

We have asked five colleagues, authorities in their several fields, to write critical commentaries on various aspects of *Prochloron*, more or less dealing with its cytology, biochemistry, physiology, ecology, and phylogeny, and these constitute chapters 3–7 of this volume. In a separate chapter, we have included some observations and hints, based largely on our own experiences in the Caroline Islands, that may prove useful to other researchers on *Prochloron* and its didemnid hosts. A few additional references, not included in the main bibliography, are given at the end of each chapter.

Over the years we have sent batches of *Prochloron* cells, variously collected and preserved, to about 80 scientists, who examined them in various ways (or plan to do so). Many of the published articles on this alga resulted from such studies. Other investigators reported to us some negative or inconclusive results that did not warrant special publications. Among these we were told of an apparent absence in our *Prochloron* samples of cyanophycin (R. D. Simon, University of Rochester, N.Y., U.S.A.), ether lipids (H. Ross, University of Leicester, England), fructose 2,6 bisphosphate (R. B. Buchanan, University of California, Berkeley, U.S.A.), NADP-isocitrate dehydrogenase (H. C. Reeves, Arizona State University, Tempe, U.S.A.), endonuclease and RNA-ligase activity (B. Javor, University of California, San Diego, U.S.A.), and nucleotide-specific DNA-ase activity (A. de Waard, Silvius Laboratory, Leiden, Netherlands). We were also told of some abnormalities in the chlorophylls and their absorption spectra (D. Doernemann, University of Marburg, Germany; J. Olie, Florida State University, Tallahassee, U.S.A.; J. Brown, Carnegie Institute of Washington, Palo Alto, Calif., U.S.A.). We are indebted to these and other colleagues for allowing us to mention their findings.

In the field of prochlorophycology new vistas opened up when Burger-Wiersma *et al.* (1986) discovered a free-living filamentous prokaryote which produces oxygen when illuminated, like a cyanophyte, but which contains chlorophylls *a* and *b* like *Prochloron* (and like the eukaryote chlorophytes). Since this alga can be easily grown in laboratory cultures, studies of its physiology, biochemistry and molecular biology may be expected to proceed rapidly. Our Dutch colleagues have kindly contributed to this book a chapter summarizing their recent work on *Prochlorothrix*, which is formally a second genus in the Prochlorophyta even if it

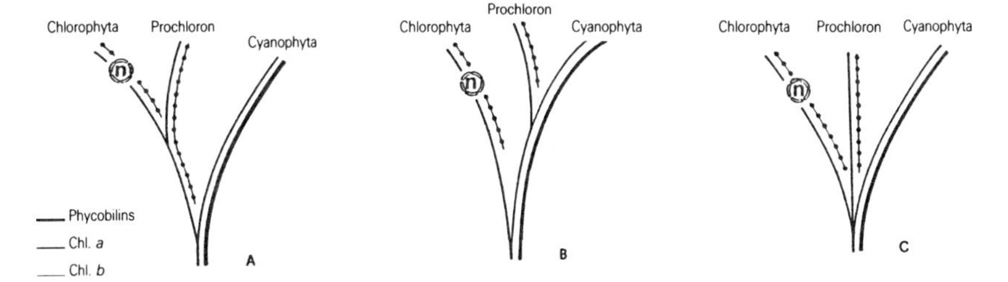

Figure 1.1 Three possible phylogenetic scenarios for origins of certain algal subclasses. (A) Certain cyanophytes lose their ability to synthesize photosynthetic bilin pigments and gain chlorophyll *b*. Among these, some then develop eukaryons and become chlorophytes; others retain a prokaryotic organization. (B) In one branch, chlorophyll *b* arises and, later, eukaryosis. In another, independent line, chlorophyll *b* arises but the cells remain prokaryotic. (C) Prochlorophytes arise independently and are no more closely related to cyanophytes than they are to chlorophytes. (Figs. 1.1–1.6 from Lewin, 1987)

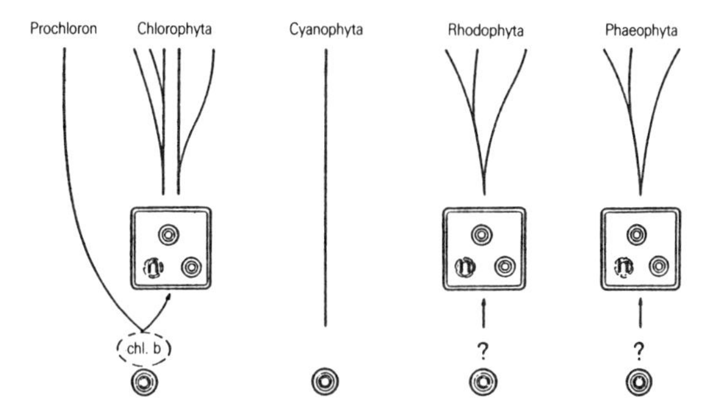

Figure 1.2 One possible phylogenetic scenario for origins of different algal subclasses. Note that here, if the origin of chlorophyll *b* is monophyletic, as shown, then one has to postulate separate and distinct origins for the eukaryon among chlorophytes, rhodophytes, phaeophytes, and so on (which seems very unlikely).

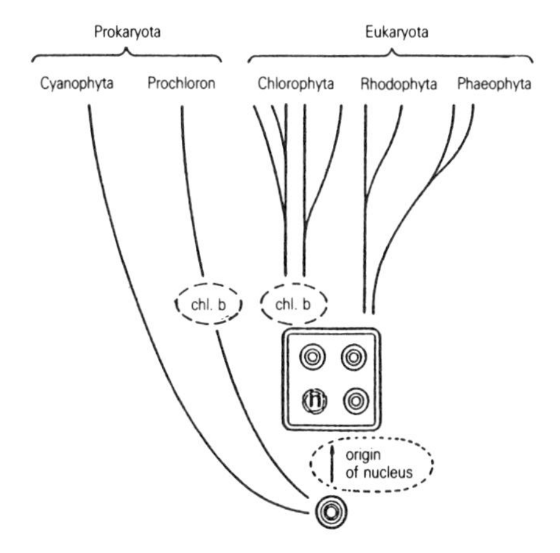

Figure 1.3 Another possible phylogenetic scenario for origins of different algal subclasses. Note that, if the origin of the eukaryon is monophyletic (which seems most probable), then chlorophyll *b* biosynthesis must have arisen separately in more than one line.

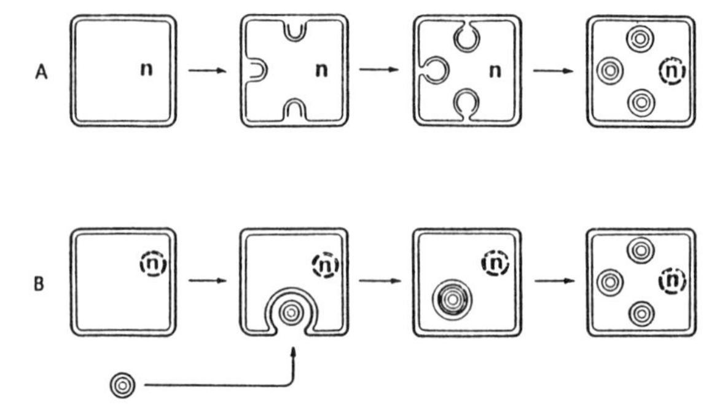

Figure 1.4 Possible scenarios for origins of chloroplasts: (A) by invagination of portions of protoplasmic membrane bearing photosynthetic pigments (i.e., direct compartmentalization or filiation) or (B) by ingestion (but not digestion) followed by intracellular reproduction of a photosynthetic prokaryote (i.e., symbiogenesis).

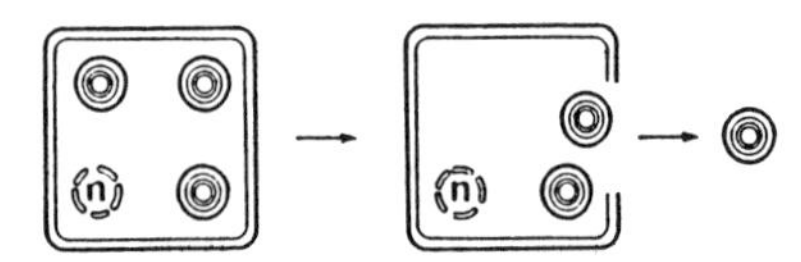

Figure 1.5 Diagram illustrating possible (but unlikely) origin of *Prochloron* from a chloroplast that is released from its parental (chlorophyte) cell and somehow achieves reproductive autonomy.

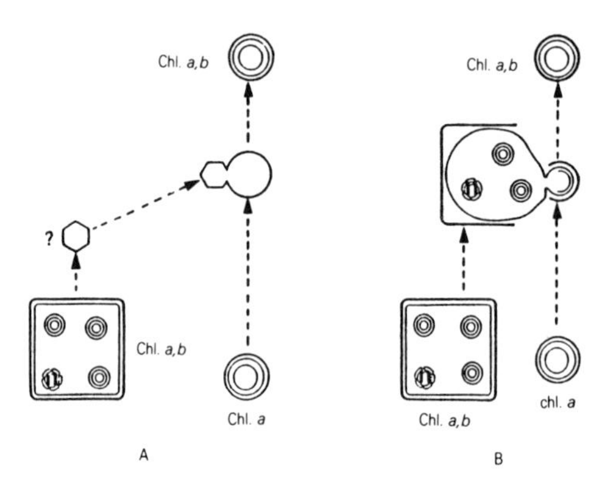

Figure 1.6 Diagrams illustrating possible scenarios whereby the ability to synthesize chlorophyll *b* could be transferred from a chlorophyte (with chlorophyll *a* and *b*) to a prokaryotic alga with chl. *a*. (A) Transmission by a putative viral vector. (B) Transmission by cell–cell fusion, perhaps in the gut of an animal.

may ultimately prove to be only distantly related to *Prochloron* (since the ability to synthesize chlorophyll *b* is probably polyphyletic; Lewin, 1986).

An even more recent discovery is that of Chisholm et al. (1988), who have reported what may be free-living prochlorophytes, in numbers as high as $10^8\ 1^{-1}$, in the dim depths of the oceans. The cells, less than 1 micrometer in diameter, apparently contain chlorophyll *b* and a pigment resembling divinyl-chlorophyll *a* in equal amounts, along with alpha carotene and zeaxanthin, and are actively photosynthetic. They were detected by flow cytometry, but have not yet been grown in culture, so provisionally they are nameless. Nevertheless, they offer exciting prospects for both comparative microbiology and—in view of their evident abundance—marine ecology.

Among the main lessons that we learned along the way was the value of scientific collaboration. So many scientific techniques have been developed in the last few years that it is quite impossible for anyone to master more than a few, even if he had the financial and technical resources to do so. Nowadays much biological research is done by teams, and, even so, the teams have to specialize: interdisciplinary, inter-institutional and international collaboration is now essential for progress. We humbly admit that we, the editors, have done very little of the critical work that constitutes the bulk of this book. Mostly, we just collected materials and solicited help from experts in working them up. Without such help, the alga we call *Prochloron* would probably still be an unnamed green smudge on obscure animals in remote locations. But with it, we have been able to compile a small new chapter of plant science.

References (Additional to Bibliography)

Chisholm, S. W., Olson, R. J., Zettler, E. R., Goericke, R., Waterbury, J. B., and Welschmeyer, N. A. 1988. A novel free-living prochlorophyte abundant in the oceanic euphotic zone. Nature 334:340–343.

Kott, P. Related species of *Trididemnum* in symbiosis with Cyanophyta. Proc. Linn. Soc. N.S.W. 107:515–520 (1984).

Monniot, F. Ascidies littorales de Guadaloupe. VIII. Questions de systématique évolutive posées par les Didemnidae. Bull. Mus. natn. Hist. Nat. Paris, 4e sér. 6A, 885–905 (1984).

Olson, R. R. Protoadaptations of the Caribbean colonial ascidian–cyanophyte symbiosis *Trididemnum solidum*. Biol. Bull. 170:62–74 (1986).

Parry, D. L. Cyanophytes with R-phycoerythrins in association with seven species of ascidians from the Great Barrier Reef. Phycologia 23:503–505 (1984).

Ruetzler, K. An unusual bluegreen alga symbiotic with two new species of *Ulosa* (Porifera: Hymeniacidonidae) from Carrie Bow Cay, Belize. P.S.Z.N.I.: Mar. Ecol. 2, 35–50 (1981).

Sybesma, J., R. C. van Duyl, and R. P. M. Bak. The ecology of the tropical compound ascidian *Trididemnum solidum*. III. Symbiotic association with unicellular algae. Mar. Ecol. Prog. Ser. 6:53–59 (1981).

Chapter 2

Collection and Handling of *Prochloron* and Its Ascidian Hosts

Ralph A. Lewin and Lanna Cheng

Introduction

Though its unique features were recognized only recently, *Prochloron*—the only marine prokaryote that photosynthesizes by using a chlorophyll *a+b* system like that of land plants—is not a rare alga on many tropical seashores. Like the zooxanthellae of corals, it exists as a symbiont in zones where other algae are sparse or absent. We have found it especially abundant in low-littoral to sublittoral zones around Palau, in the West Caroline Islands, where at least six species of didemnids (colonial ascidians), occupying dissimilar and distinct ecological niches, harbor this prochlorophyte symbiont. Massed colonies of one of these hosts, *Tridi-demnum cyclops*, cover a large proportion of the exposed surfaces 1 or 2 m below mean sea level off Kamori island, in a swath several meters wide and several hundred meters long; its biomass may be conservatively estimated at 0.1 tons (wet weight), of which at least 10 kg is represented by *Prochloron*. Per unit area, the chlorophyll *a* and *b* contents of such areas of symbiotic didemnids, and presumably therefore the productivity of the *Prochloron*, are comparable to those of, say, a field of grass or the phytoplankton in a column of reasonably fertile water.

We are grateful to colleagues who joined us in various *Prochloron* expeditions, and especially to Dr. Randall S. Alberte, for many useful suggestions and contributions to this chapter.

Despite its ecological success in warm marine waters, *Prochloron* has not proved amenable to laboratory culture, and all that we now know about it has been based on material collected from nature. In the hope of encouraging and facilitating further studies of this interesting organism, we have compiled here a few observations and hints, based on our own experiences in the past 15 years on Palau and other tropical islands, on the collection and handling of this symbiotic alga and the didemnid ascidians with which it is normally associated.

Prochloron is a tropical alga, regularly found in association with certain didemnids in coastal waters where the sea temperatures range from 21° to 31°C. It occasionally occurs as an epizoic film on outer surfaces of other species of *Didemnum* (e.g. *Did. candidum* Savigny, 1816) in Baja California, Mexico (Plates 2–26, 2–28, 2–32) and, rarely, on other marine invertebrates (Cheng and Lewin, 1984; Kott et al., 1984; Parry, 1986) but not regularly inside the colonies. Such epizoic cells are generally smaller (8–14 μm in diameter) than endozoic ones (12–25 μm). They can be washed free from the host, and with some trouble one can collect milligram quantities, but these are two or three orders of magnitude less than the amounts easily expressible from *Lissoclinum patella* (see later discussion), and very much more heavily contaminated with diatoms and other organisms from the didemnid surface.

The taxonomy of didemnids is not simple, but with a little experience some of the commoner symbiotic types can be recognized even by nonspecialists. Major types are illustrated in Plate 2.1, and some of their physical features, indicating their relative convenience as sources of *Prochloron* cells, are summarized in Table 2.1.

We are indebted to Drs. P. Kott (Queensland Museum, Brisbane, Australia), F. Lafargue (Laboratoire Arago, Banyuls, France), and C. and F. Monniot (Museum Nationale d'Histoire Naturelle, Paris, France) for help with identifications of the various host species. We recommend that researchers keep samples, preserved with formalin, for reference in case questions arise about the specific identities of the animals with which they have worked.

Collection and Handling

Didemnum molle. Perhaps the commonest and most widespread of obligately symbiotic didemnids is *Didemnum molle* (Herdman, 1886), the colonies of which are more or less isodiametric with about the size and shape of acorns or plums (Plates 2–17 through 2–24). They occur on all kinds of substrates: dead coral, ropes, pipes, shells, even a silty bottom. Sometimes the colonies are densely packed in masses of 10–100, but

Table 2.1 Some physical characteristics of didemnid colonies and their associated algae (*Prochloron* sp.) as found on coasts of Palau, W.C.I.

Host (ascidian)	Common Colony Dimensions (mm)			Spiculo spheres	Mucus	pH of Expressed Fluid	Ease of Expression of *Prochloron*	Diameter of *Prochloron* cells (μm)
	Length	Width	Thickness					
Didemnum molle	10–20	10–30	10–30	+	++++	6–7	(+)	16–28
Diplosoma similis	10–50	10–20	2–5	–	–	1–2	+++	9–18
Dip. virens	10–20	5–15	2–4	–	–	3–6	+	9–14
Lissoclinum patella	30–200	25–150	10–15	+	–	3–5	+++	13–22
Liss. Punctatum	10–20	5–10	3–5	+	++	3–4	(+)	18–30
Liss. voeltzkowi	10–50	10–20	3–5	+	–	6–7	+	9–13
Tridídemnum cyclops	5–10	5–10	3–5	+++	–	1–2	++	9–18

often they occur singly. They vary widely in color—white, brown, olive-green, slate, dull violet, and so on. Usually in one site all colonies are concolorous, suggesting that they are clonal. Undisturbed, colonies of *Did. molle* are inflated, with usually a single, widely distended cloacal orifice, inside which one can see the *Prochloron* cells packed as a greenish black mass. When touched, or disturbed by moving water, the animals contract in a second or two to about half-volume, and the orifice closes to a slit. Intact colonies can be detached if care is exercised, but they tear very easily. They can survive for a few days in tanks with running seawater, but in less favorable conditions they tend to die and decay all too readily.

We know of no way whereby algal cell preparations can be obtained from *Did. molle,* because the cells are embedded in masses of a mucopolysaccharide (containing about 10 percent protein, 20 percent uronic acid, and 10 percent sulphate; J. A. Christiansen, personal communication), from which they cannot be separated by centrifugation or any other method that we have tried. Therefore, this species, although among the commonest, is also among the least satisfactory as a source of symbiotic algae. This is a pity, because the *Prochloron* cells in *Did. molle* differ in several ways from those of other didemnids. They tend to be larger, up to 30 μm in diameter—an appreciable size for a prokaryote! They contain not one but several vacuoles, of unknown content. Other differences, including unique features of fine structure, are described by Swift in chapter 7.

Diplosoma virens. *Diplosoma virens* (Hartmeyer, 1909) is the most tolerant of dessication, at least in our experience in Hawaii and on Palau. It grows, often as sheets of clustered colonies, on surfaces of mangrove roots and branches that may be exposed for several hours at low-tide periods (Plates 2–8, 2–13, 2–15, 2–16). It has been observed to survive for many months on the running-seawater table at Coconut Island, Hawaii. The colonies, which are usually dark green (rarely pinkish or blue), lack the calcareous spiculospheres of other didemnid genera. They spread by slow colony migration at rates up to 1 cm per day, by growth and by colony subdivision, and they can be readily detached from the substratum. They have been maintained for a few days in small aquaria with less than continuous aeration. The algal cells are mostly embedded in the test (the gelatinous, nonliving matrix of the colony), from which they cannot readily be separated by any mechanical means that we have tried. However, small quantities of cells have been collected by gently tickling colonies with the tip of a squeeze-bottle (Withers et al., 1978).

Diplosoma similis. Colonies of *Diplosoma similis* (Sluiter, 1909) occur on dead coral (*Acropora*), usually in shaded notches, and are quite common in some areas (e.g., by fringing reef drop-offs at Palau, the Solomon Islands, and among coral rubble at Phuket, Thailand). The colonies are irregular, 1–10 cm long or longer, dark green, smooth, and relatively fragile (Plate 2–12). One cannot easily detach them from the substrate without tearing the colonial surface and releasing many or most of the algal cells, which in this species occur in large confluent lacunae. At the same time, the damaged animal tissue releases acid (presumably sulphuric), which may cause the pH of the seawater in small containers to fall below 2. Unless carefully buffered, suspensions of *Prochloron* cells from this (and certain other) species of didemnid become so acidic that the cells soon die and discolor, to olivaceous brown, presumably because of phaeophytinization. If such difficulties could be overcome, however, *Dip. similis* might prove a good source of free *Prochloron* cells, since in this species they constitute more than half of its total biomass and are not embedded in test or mucus material.

Trididemnum cyclops. *Trididemnum cyclops* (Michaelsen, 1921) occurs on all kinds of submerged surfaces, such as those of dead coral, shells, and dead leaves (Plate 2–7, 2–14, 2–25). The colonies are usually greyish-green above and white below, the underside tissues containing an abundance of calcareous spiculospheres. They are usually only 2–10 mm in diameter; have a firm, limy consistency; and can be readily detached. *Prochloron* cells can be expressed from their cloacal systems, but the yields are not large (some 5,000 to 20,000 per colony). Here, too, care must be taken to ensure that acidification, from bruised animals, does not kill the algal cells. Adequate buffering (to pH 7.5–8.3) can be achieved with 40 m*M* Tris, 40–100 m*M* Bicine, or 10–20 m*M* sodium bicarbonate, which are not toxic (see Alberte et al., 1986).

Lissoclinum punctatum. *L. punctatum* (Kott, 1977) has somewhat gelatinous, deep green colonies, rather like small frogspawn, that occur on lumps of coralline algae (e.g., *Lithothamnium*) out of the range of direct sunlight, usually just under overhanging edges (Plate 2–5, 2–11). It is of particular interest since a large proportion of its *Prochloron* cells are individually enclosed in a thin layer of nucleated host cell protoplasm, indicating either ingestion (as suggested by Cox, 1983, although we have noted no evidence for digestion of the enclosed algal cells) or a closely integrated kind of symbiosis. The species is uncommon around the Kamori area in Palau, where many other species of symbiotic didemnids abound, but is abundant on some of the less accessible western islets. It has a further disadvantage for the experimenter because it is so readily subject to autolysis when the colonies are mechanically damaged.

Lissoclinum voeltzkowi. *Lissoclinum voeltzkowi* (Michaelson, 1920) is common and plays an important role in the ecology of sea-grasses (*Halophila* sp. and *Enhalus* sp.) and siphonaceous algae (*Halimeda* spp.), covering large areas of the leaves and thus presumably impairing their photosynthetic activity (Plate 2–9, 2–10). The colonies are gray-green, with pale edges. They are about 1 mm thick and up to 30 cm long, though usually only in the range of 1–10 cm. They can easily be peeled off from their substrates without suffering damage. In illuminated tanks with running seawater, they readily proliferate by fragmentation and tend to colonize tank edges at the water surface. The *Prochloron* cells of *L. voeltzkowi* are largely embedded in the test, from which they cannot easily be expressed. Pressed colonies yield not only algal cells but also a brilliant yellow, water-soluble pigment, as yet unidentified. This metabolite may be responsible for the somewhat astringent taste of *L. voeltzkowi* and act to deter browsing by fishes. (Incidentally, we have noted no animals browsing on any symbiotic didemnids, presumably because of the distasteful chemicals, acids, mucus, or spiculospheres that characterize the various species.)

Lissoclinum patella. For many purposes the most convenient source of *Prochloron* cells is the giant didemnid, *Lissoclinum patella* (Gottschaldt, 1898). Single colonies are gray-green, with raised ridges 2–3 mm wide; they may reach 10–25 cm in diameter, be 1–2 cm thick, and weigh up to 200 g (Plates 2–1, 2–2). Kept submerged in running seawater tanks at 28°–30°C, they survive well and remain apparently healthy for weeks, though if held for more than a few minutes out of water, or stored with inadequate aeration under stagnant conditions, they become oedmatous in a few hours, soon die, and quickly rot (and stink!). For physiological experiments on the symbiotic system we have found it convenient to use colonies of uniform size, which we make by punching disks out of large, more or less flat colonies (Plates 2–3, 2–4). Using an 18-mm-diameter cork borer, carefully sharpened, we have been able to make many dozens of such disks. Their edges heal in two days, and if needed each develops a new cloacal aperture. When attached to a substratum the disk may remain discoid or flattened; others, unattached, tend to become more or less spherical. Survival is almost 100 percent, and photosynthetic and respiratory activities appear to be normal (Alberte et al., 1987).

Although we have never successfully established living and growing didemnid colonies in an aquarium away from the original location, we suggest that the following procedures may ultimately prove successful. Small, intact, healthy colonies, freshly collected, might stand the best chances of surviving air transportation. They should be kept in a relatively large volume of seawater, and not subjected to cold shock (below

25°C) for prolonged periods of time. To avoid the low temperatures of air-conditioned planes and airport terminals, they should be kept warm in vacuum flasks; one should check their temperature frequently and rewarm the water when necessary. Constant aeration can be provided with a portable battery-operated air pump. Alternatively, specimens could be enclosed, with plenty of water, in strong polyethylene bags, gassed with oxygen, tightly sealed, and shipped in an insulated (polystyrene foam) box, preferably of a size and shape to slip under a plane seat, so that it can be taken on board and tended by the investigator. Potentially adverse conditions in cargo holds could thereby be avoided. We brought back from Palau a few small colonies of *L. patella* and managed to keep one alive for a week in an aerated flask, illuminated by diffuse daylight and maintained at ca. 26°C, in which we changed the water daily. Then it died, possibly of starvation. Almost certainly even symbiotic didemnids have to ingest some phytoplankton, unless the algae can supply both organic carbon and the amino acids essential for growth; but we do not know what food organisms would be suitable.

The especially valuable feature of *L. patella* is the ease with which its colonies yield pure suspensions of *Prochloron* cells when manually pressed. Before being pressed, the undersides of the colonies should be picked free from bits of shell, segments of *Halimeda*, and so on. They are then folded into a sheet of Nitex nylon mesh, ca. 180 μm in pore diameter, which retains most residue debris. No cutting is necessary: 95 percent of the algae can be easily pressed out through natural apertures, leaving the colonies pale flesh-pink (and, incidentally, dead). Acid liberated by the bruised host colonies can reduce the pH of seawater to 6 or 5, levels lethal to the algae. For this reason, we customarily express colonies (about 100 g) into 500 or 1,000 ml of seawater buffered with 40–100 m*M* Bicine or 10–20 m*M* sodium bicarbonate, checking to ensure that the pH remains above 7.2. Suspensions of algal cells prepared in this way are initially almost free of contaminant microbes (Plates 2–31), suggesting that in this symbiosis the host selectively favors the sole growth of *Prochloron*. However, in such suspensions, even in well-buffered seawater, the cells tend to degenerate because of at least three adverse factors, as listed below.

Treatment of *Prochloron* Cells

a. *Prochloron* cells, on lysis, liberate phenolic substances (Barclay et al., 1987), which are undoubtedly toxic. (These compounds have been responsible for difficulties experienced in demonstrating various enzyme activities in frozen or freeze-dried preparations of this alga.) To reduce the adverse effects of these phenolic compounds, liberated on lysis of dead cells, we have found it beneficial to add about 1 g/liter (w/v) polyvinyl polypyrrholidone to the expression medium, except when the presence of this insoluble agent might interfere with subsequent handling of the cells.

b. Soluble proteins from the didemnid blood provide substrates for the growth of heterotrophic bacteria, which rapidly proliferate (especially at tropical temperatures) and soon may render unstirred suspensions anoxic and putrid.

c. Protozoa also proliferate, eating both bacteria and algal cells.

To some extent, contaminants (b and c, just cited) can be reduced by two or three sequential washings of the algal cells, which are relatively heavy. In 10 minutes most sink from suspension to the bottom of a tube or beaker; centrifugation in a clinical centrifuge for as little as 5–10 s deposits all the cells in a pellet. (We have found that longer centrifugation is not advisable, since it may result in cell damage, perhaps due to anaerobiosis.) They readily aggregate into clumps: for reasons we do not understand, clumping is reduced when the cells die (as well as in media containing a chelating agent like EDTA). The relative heaviness of the living cells is presumably related to their normally spherical shape and high optical refractility; they are evidently very turgid because of hypertonicity of the cell sap. When *Prochloron* cells die they lose refractility, and their diameters decrease (Lewin and Cheng, unpublished) as the cell wall pressure falls to zero. (Living cells could perhaps be separated from dead ones by gravity or gradient centrifugation in a noninhibitory solute system.)

A further indication of cell death is the rapid coagulation of the cell contents. Pressed on a microscope slide under a coverslip, dead cells break open to release a coagulated blob of protoplasm, whereas live cells release a viscous green fluid. (In the "English press" technique, a small droplet of suspension is mounted under a coverslip, surplus seawater is blotted away by gentle pressure under absorbent paper, and then a patch of cells, selected in view under 100× magnification, is pressed firmly with the tip of a hard pencil, knife point, or dissecting needle.)

Using these criteria, one can fairly easily distinguish living from dead cells of *Prochloron*, and it is now clear that in most of our earlier efforts to grow isolated cells, for instance, we had been working with cells that had already died. *Prochloron* cells are especially sensitive to cooling and should not be handled in air-conditioned laboratories. At 25°C their photosynthetic activity is only one half of that at 30°C; at 20°C it falls almost to zero (Alberte et al., 1986). Cells cooled for a few minutes (e.g., by immersion of a suspension in iced water) have contents that are demonstrably congealed and presumably dead. Incidentally, cells killed in this way release cell sap with a refractive index two or three times that of seawater, the high value being presumably attributable to a high salt content. We have found that cells of *Prochloron* survive as well, or better, in seawater supplemented with an additional 0.3 *M* of salt (sodium chloride or sulphate) as in normal seawater. (We do not yet know whether in the host they experience comparably elevated osmolarity or, if so, to what extent the hypertonicity of the cell sap may be related to the symbiotic association.) For the preparation of cells for electron microscopy it is probably best to fix them in a medium hypertonic to seawater, closer to the osmotic pressure of the cell contents. We suspect that the inflation of thylakoids seen in many of the published transmission electron micrographs of *Prochloron* may be attributable to osmotic distension.

Conclusion

So far, the cultivation of *Prochloron*, away from its host, has baffled us. Despite hundreds, perhaps thousands, of tested combinations of mineral nutrients and vitamin supplements at various salinities, pH values, temperatures, and light regimes, we have not succeeded in establishing sustained cultures of this recalcitrant alga. Patterson and Withers (1984) reported a partial success, some three cell divisions in 15 days, although they failed to establish that growth had been really an increase in biomass, and not simply an increase in cell number associated with a corresponding decrease in cell size. They had supplemented seawater media with tryptophan. We have not even been able to repeat this limited success. The trouble with the addition of organic supplements like this, or various extracts of host colonies or sera, is that, in xenic and warm conditions, contaminant bacteria and protozoa soon overgrow, kill, and consume the *Prochloron* cells used as inoculum. And it is not easy to establish axenic conditions when we still do not know what factors this alga requires for growth. We have often tried to establish cultures from single cells, which when freshly removed from host animals can be isolated bacteria-free, but even this expedient has failed with monotonous regularity. The keys to the solution of this problem probably include correct balances of osmolarity, oxygen and CO_2, light and darkness, but we have not yet been able to hit on a winning combination.

However, much biochemical work has been done on freeze-dried material of *Prochloron* cells, expressed from *L. patella* (primarily) and concentrated by settling or gentle centrifugation, frozen to −20° and then either freeze-dried on location (we took a lyophilizer to Palau, where we used it for two or three years) or transported frozen in ice for later lyophilization in temperate-zone laboratories. Scientific data based on such material are described in the other reports presented in this volume.

Plate 1. *Lissoclinum patella,* upper surface of large colony showing ridges (from which silt has not been swept away) and cloacal aperture (left of centre). Palau. Scale, approx. × 1.4.

Plate 2. *Lissoclinum patella,* portion of colony showing individual animals' mouths as pale green dots. Palau. Scale, approx. × 1.4.

Plate 3. *Lissoclinum patella* colonies, one perforated by cork-borer to remove standard 'plugs' for experiments. Upper surfaces of colonies are whitish and wrinkled, lower surfaces are in places pinkish because of growth of a crustose red alga. Palau. Scale, approx. × 0.15.

Plate 4. *Lissoclinum patella*, healed disc cut from colony, showing individual animals as small, pale dots and one cloacal aperture as dark ring. Palau. Scale, approx. × 2.

Plate 5. *Lissoclinum(?) punctatum* (green), growing with red algae *(Amphiroa)* and *Caulerpa* (a bluish green alga). Palau. Scale, approx. × 0.7. (Photo: C. Birkeland.)

Plate 6. Unidentified peacock-blue didemnid, containing *Prochloron*. Munda, Solomon Islands. Scale, approx. × 1.4.

Plate 7. *Trididemnum cyclops* on coral. Palau. Scale, approx. × 1.4.

Plate 8. *Diplosoma virens,* blue variety, near low-water mark. Pulau Samarkau, Singapore. Scale, approx. × 1.4. (Photo: W. Vidaver.)

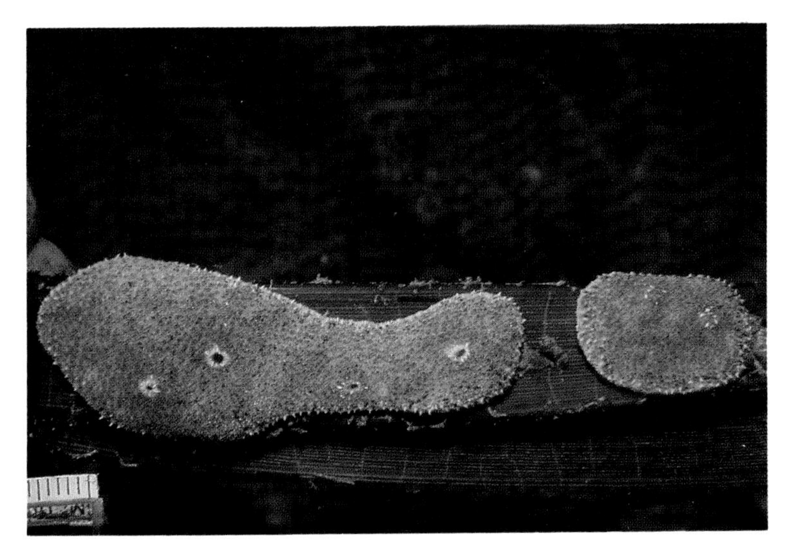

Plate 9. *Lissoclinum voeltzkowi* colonies on leaf of sea grass *(Enhalus)*. Koror, Palau. Note 1–4 cloacal apertures. Scale, approx. × 1.4.

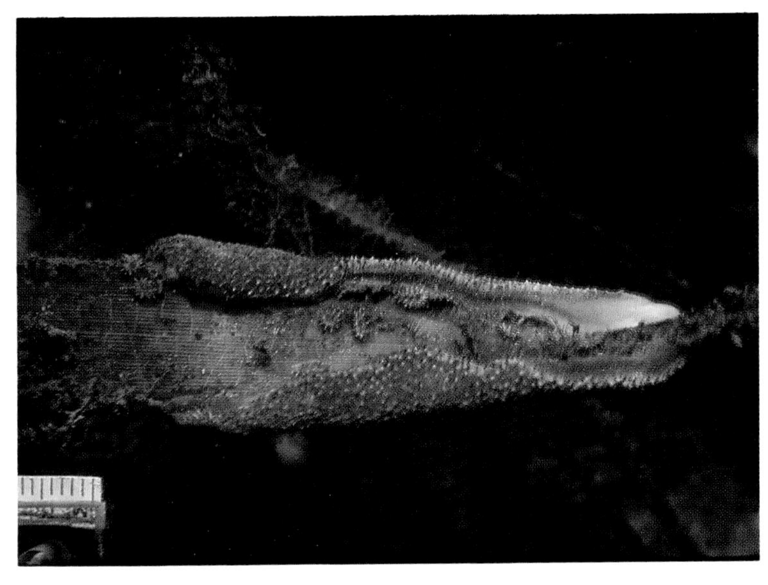

Plate 10. *Lissoclinum voeltzkowi* large colony on leaf of sea grass *(Enhalus)*, over which it has folded at both edges. Note also young colonies. Koror, Palau. Scale, approx. × 1.4.

Plate 11. *Lissoclinum punctatum* on orange-coloured sponge; Koror, Palau. Scale, approx. × 1.4.

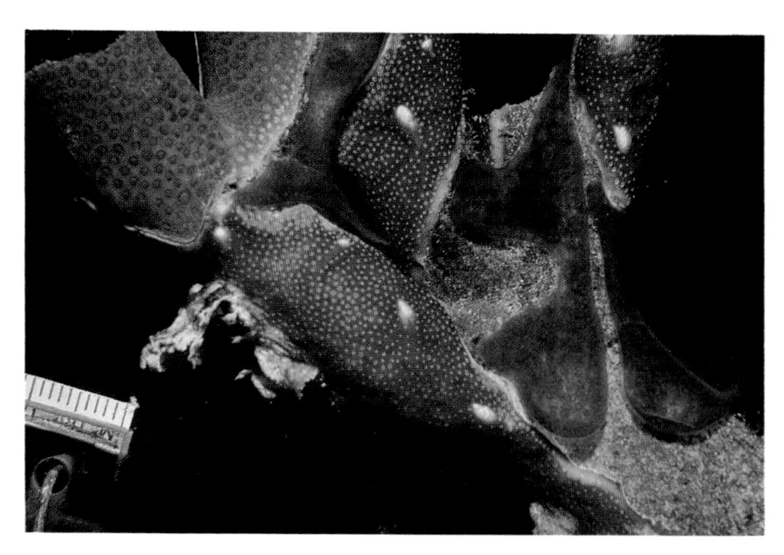

Plate 12. *Diplosoma similis* (very dark green), with a non-symbiotic didemnid (mauve); Koror, Palau. Scale, approx. × 1.4.

Plate 13. *Diplosoma virens,* two color variants (one typically green, the other with a lilac tinge); reef flat, Rock Islands, Palau. Scale, approx. × 1.4.

Plate 14. *Trididemnum cyclops* on dead coral. Koror, Palau. Scale, approx. × 1.4.

Plate 15. *Diplosoma virens* colonies on ascending rhizophores of the mangrove *Sonneratia*. Palau. Scale, approx. × 0.2. (Photo: R. L. Pardy.)

Plate 16. *Diplosoma virens* on mangrove log exposed at low tide. Koror, Palau. Scale, approx. × 0.3.

Plate 17. *Didemnum molle,* tan-colored colonies. Philippine Is. (Photo: Twila Bratcher.) Scale, approx. × 1.

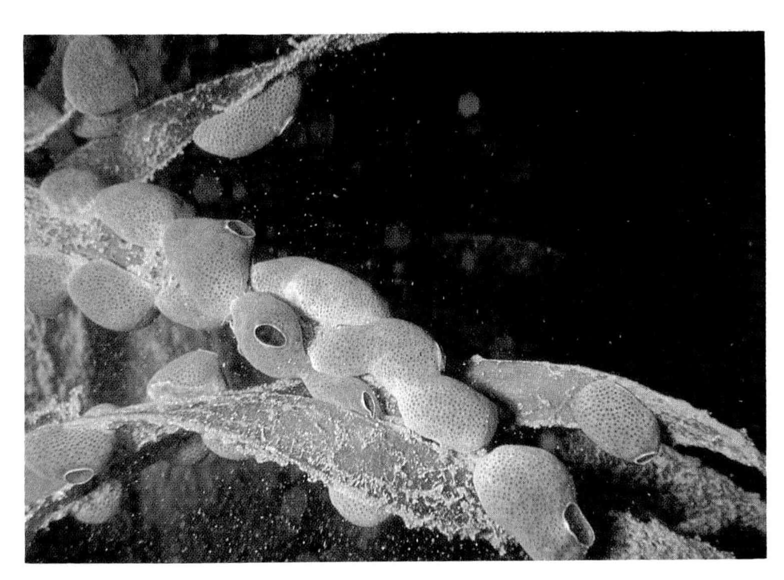

Plate 18. *Didemnum molle,* olive colonies, on leaves of sea grass *(Enhalus);* Koror, Palau. Scale, approx. × 1. (Photo: F. Barnwell.)

Plate 19. *Didemnum molle*, young colonies among yellow sea-cucumbers. Philippine Is. Scale, approx. × 1. (Photo: R. Yin.)

Plate 20. *Didemnum molle*, young colonies, with non-symbiotic didemnid (pink) and various algae (*Halimeda*, *Galaxaura*, etc.) Palau. Scale, approx. × 1. (Photo: H. Hahn.)

Plate 21. *Didemnum molle,* brown and white colonies, with one cut open to show *Prochloron* cells embedded in mucus. Palau. Scale, approx. × 1.4. (Photo: L. Muscatine.)

Plate 22. *Didemnum molle,* white colony with green cloacae (*Prochloron* emergent) and brown colony with white cloacae, among corals, etc. Shrivelled texture and more or less closed cloacae indicate that the colonies have been recently disturbed. Note white attachment filaments at base of colony at bottom right. Koror, Palau. Scale, approx. × 1.4.

Plate 23. *Didemnum molle,* olivaceous colonies on coral. Dilated cloacal apertures indicate that animals have not been disturbed. Palau. Scale, approx. × 0.7. (Photo: H. Hahn.)

Plate 24. *Didemnum molle,* brown colonies, two in division stages, with distended green cloacae, on dead coral. Palau. Scale, approx. 1.5. (Photo: L. Muscatine.)

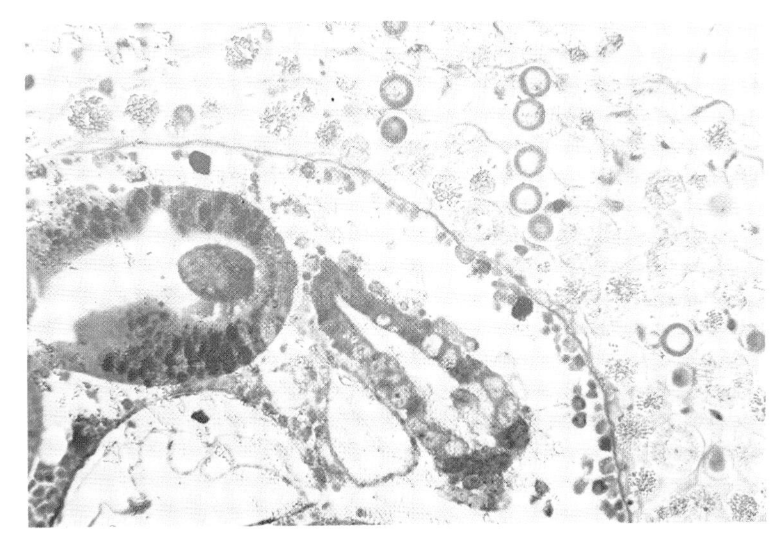

Plate 25. *Trididemnum cyclops:* section through colony lightly stained with Toluidine blue, showing *Prochloron* cells and calcareous spiculospheres embedded in "test" (organic matrix of colony). Note that, to the left, the animal cells are free from algae. Scale, approx. × 400. (Section: N. D. Holland.)

Plate 26. *Prochloron* growing as a thin layer in green patches on the holothurian *Synaptula lamperti* Heding, on a sponge just below low-water mark. Koror, Palau. Scale, approx. × 1.4.

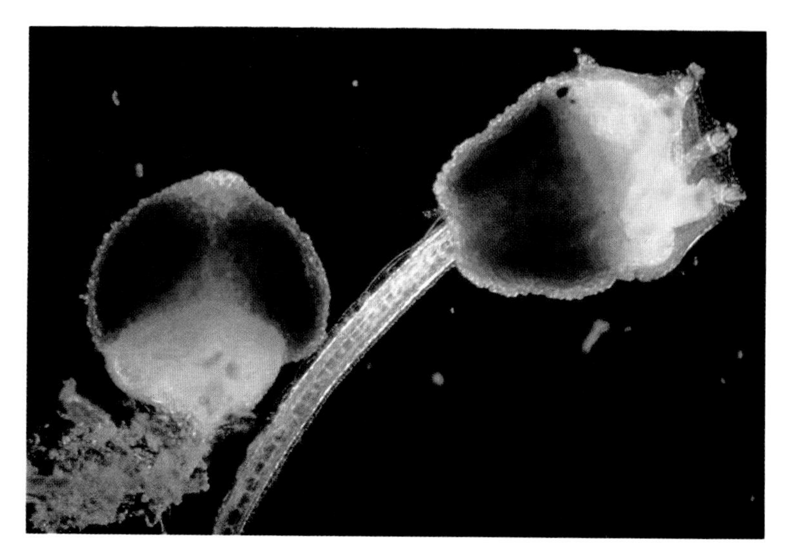

Plate 27. *Lissoclinum patella* tadpoles, one with tail retracted in early stage of maturation to form a young colony. Note latter ⅔ of body enclosed in a layer of *Prochloron* cells. Lizard Island, Qld., Australia. (Photo: Oxford Scientific Films.) Scale, approx. × 20.

Plate 28. *Didemnum candidum*, a normally non-symbiotic species, with *Prochloron* cells forming a green layer on the upper surface. Galápagos Islands, Ecuador. Scale, approx. × 1.4.

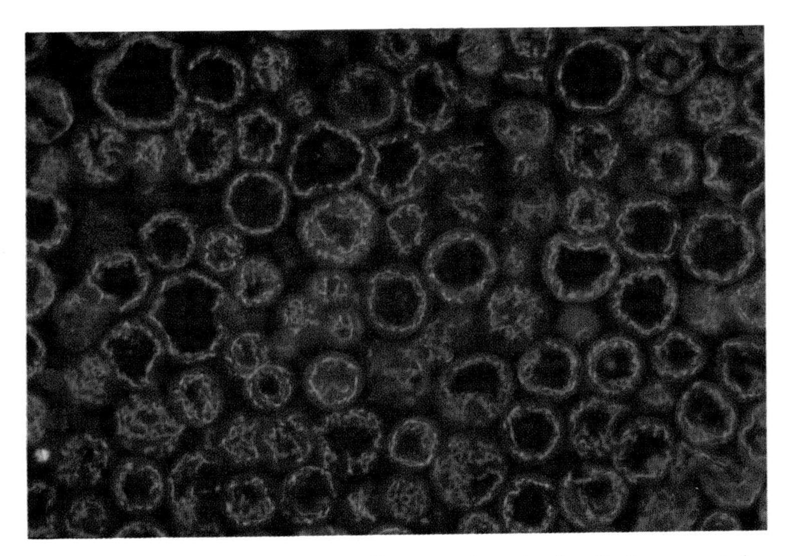

Plate 29. *Prochloron* cells from *Lissoclinum patella* stained with fluorescent dye (diamidino-phenylindole = DAPI), showing nucleic acid bodies in peripheral layer. Scale, approx. × 800. (Photo: H. Swift.)

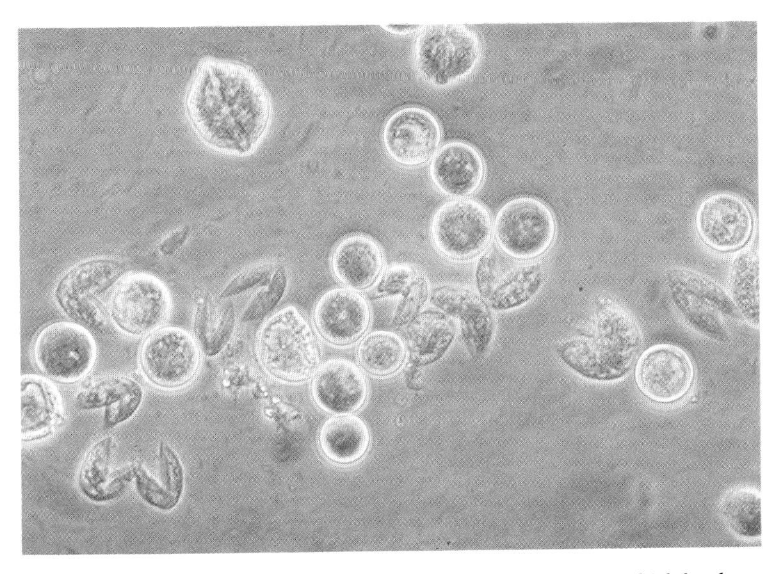

Plate 30. *Prochloron* cells from *Lissoclinum patella,* in a preparation which has been gently crushed to show how cell walls break open into neat halves. Scale, approx. × 800. (Photo: H. Swift.)

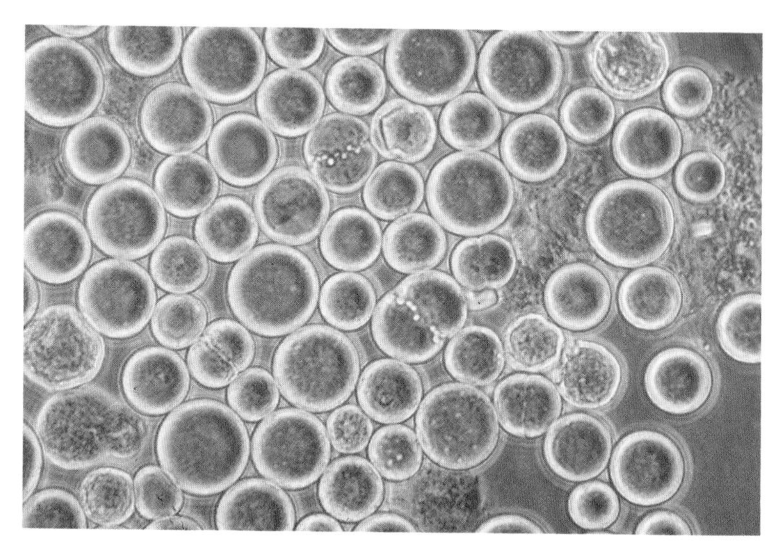

Plate 31. *Prochloron* cells from *Lissoclinum patella*. Note living cells are high refractile, whereas dead (?) cells are less refractile and somewhat shrunken. Some stages of cell division show formation of droplets across dividing face. Scale, approx. × 800. (Photo: H. Swift.)

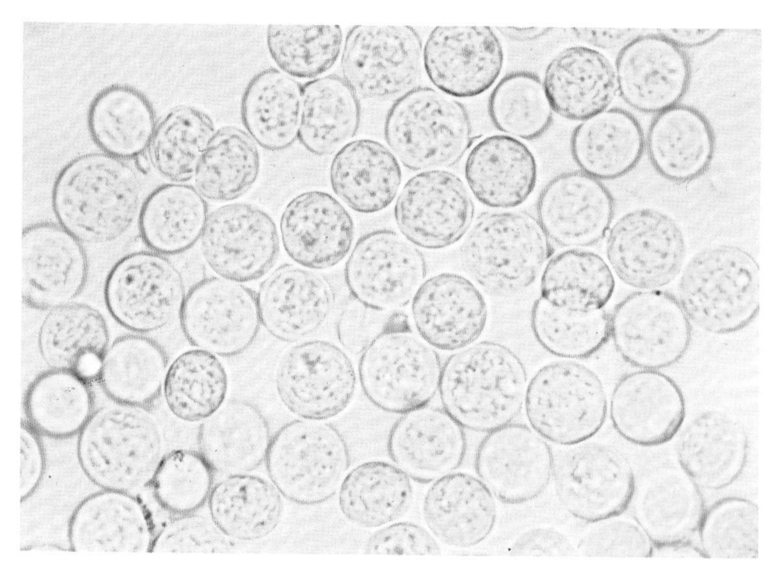

Plate 32. *Prochloron* cells from surface of *Didemnum candidum*, Puerto Peñasco, B.C., Mexico. Scale, approx. × 800. (Photo: R. Hoshaw.)

Chapter 3

Prochloron in Symbiosis

Rosevelt L. Pardy

Introduction

This section is devoted to papers mainly concerning the biology of the intact *Prochloron*–ascidian symbiosis. The term *symbiosis* has become imprecise (Pardy, 1983). I cannot resist commenting on an example from the *Prochloron* literature to illustrate this point. In the first paragraph of an interesting and important note, the author uses the term *nonobligate symbiosis*, but later refers repeatedly to *nonobligate association* until the last sentence of the article, in which both terms are used. Now although *nonobligate symbiosis* suggests that each of the partners can make do without the other, the data indicate that one organism (*Prochloron*) is promiscuous, whereas the other (the ascidian) may be indifferent. To use the term *symbiosis* for such a situation is to risk obscuring a potentially important biological phenomenon: the apparent affinity of *Prochloron* for certain mucus-laden animals. Perhaps the author sensed this, hence the apparently unconscious but still ambiguous terminology switch. I use the term *symbiosis* operationally, in a restricted sense, simply to indicate the case where *Prochloron* cells live on or in certain ascidians and a couple of other kinds of invertebrates. I think this approximates the original meaning of the term as coined by deBary. Some workers (Lewin, 1982) and many people outside of the field consider symbiosis synonymous with mutualism. This is unfortunate because they thereby endanger an objective analysis of the association. It might be better to assume that *Prochloron* occupies (or defines) a specific ecological niche and concentrate on discovering the factors differentiating and restricting the niche. Then the more interesting question arises: Does the *Prochloron*–ascidian symbiosis occupy or define a niche otherwise unavailable to the constituents when not in symbiosis?

As elsewhere (Pardy, 1983), I call animal–algae associations "phycozoans" (seaweed–animals). In a phycozoan, the algal component is called the phycobiont (a term borrowed from lichenology); the corresponding animal component is called the zoobiont. Thus, in the present case, *Prochloron* cells are the phycobionts, and the ascidian is the zoobiont. I find *phycozoan* a useful term because it doesn't require qualifying adjectives and disclaimers, or need extensive "verbal handwaving" to adjust its meaning.

Because of the interest aroused by its unusual features, the *Prochloron* component has received a greater degree of attention than the ascidian component of the phycozoan. Although several good descriptive, taxonomic, and biogeographical studies have been published, until comparatively recently there has been a lack of published physiological and biochemical studies to define *Prochloron*'s niche on or in the host. One serious handicap to cell biology studies of *Prochloron* has been our inability to maintain the cells *in vitro* in long-term culture on a defined medium. In this regard, empirical approaches to medium formulation seem to have been unsuccessful. Information concerning the ascidian–*Prochloron* interface, at the physiological and biochemical level, might be helpful for developing a more rational approach to sustained *Prochloron* culture, as well as for improving our understanding of the biology of this unusual organism.

Over the years biologists have studied alga–invertebrate symbioses in a variety of ways, attempting to answer some rather specific questions. I have recast some of these questions to use as a framework for commenting on some *Prochloron* papers. I hope this approach will stimulate further studies of the symbiosis.

Photosynthesis

Do the phycobionts photosynthesize *in hospite*, and, if so, is there net productivity?
What is the magnitude of the phycobionts' contribution to the host's metabolism?
What are the photosynthetic properties of the phycobionts?

The usual experimental approach used by workers to study these questions involves direct measurement of the oxygen exchange between the phycozoan and the surrounding environment. Typically, whole phycozoans are placed in vessels of water maintained under controlled conditions. Changes in the oxygen content of the water are determined either periodically by titration of water samples (Winkler method) or continuously by means of an oxygen electrode. In either case, comparisons are made between phycozoans maintained in the light, with corresponding controls kept in darkness. The steady production of oxygen in the light is taken as evidence for photosynthetic activity of the phycobionts. In a novel approach, Alberte et al. (1987) used standard disks cut from flat didemnid colonies with a cork borer. The disks were small enough to fit into an electrode chamber and allowed to heal before experiments were performed.

Although technically the methodology is fairly straightforward, there are special considerations when interpreting the data. For example, if the phycobionts are undergoing net photosynthesis, oxygen should be evolved. However, because of their close physical proximity, the simultaneous respiratory uptake of oxygen by both phycobionts and zoobiont leads to a significant underestimate of the amount of oxygen evolved during photosynthesis. Hence corrections for biont respiration need to be made. These involve measuring respiration rates under nonphotosynthesizing conditions, such as in the dark. This is an operational solution to the problem of separating oxygen uptake from oxygen consumption in the phycozoan. However, there remains the uncertainty as to whether respiration rates of either biont are affected by light and/or photosynthesis. Presently there is no completely satisfactory solution to this problem.

With appropriate controls and guarded assumptions, oxygen evolution measurements can be very useful for assessing photosynthetic rates, and workers typically probe a range of environmental factors for their effect(s) on photosynthesis. Among these factors, light intensity (photon flux density) is perhaps one of the most important. Typically photosynthetic rates are determined under a range of experimentally controlled light intensities. From these data three important indices are derived. I_k is the relatively high light intensity at which maximum photosynthesis occurs, where the photochemical reaction centers are assumed to be saturated with photons. Hence, I_k is frequently referred to as the light saturation level. I_c is the relatively low light intensity at which photosynthetic oxygen production just equals respiratory oxygen uptake. This is called the compensation point. Finally, the initial slope of the photosynthetic rate–light intensity curve indicates the relative sensitivity of the photochemical systems to changes in light intensity. The absolute values of these indices are ultimately determined by the size and number of photochemical reaction centers in the photosynthesizing cells. Of scientific interest, and of ecological importance to the phycobionts and ultimately to the phycozoan, is the fact that the number and configuration of the photosynthetic reaction centers are affected by the light intensity regime under which the phycobionts are maintained. Thus, the I_k and I_c are useful indicators of the light environment of the phycozoan (and *vice versa*).

Papers by Tokioka (1942), Thinh and Griffiths (1977), Sybesma et al. (1981), Pardy (1984), Olson and Porter (1985), and Alberte et al. (1986, 1987) describe research involving oxygen exchange measurements of intact phycozoans. Though he mistook the algae for zoochlorellae, Tokioka (1942), using the Winkler method, was the first to report oxygen production by intact *Prochloron*–ascidian phycozoans in Belau (Palau). In fact, so vigorous was the oxygen evolution in his experiments that some of the ascidians were carried to the top of his bottles by oxygen bubbles.

Olson and Porter (1985) presented a comprehensive study of photosynthesis by an ascidian–*Prochloron* phycozoan (*Didemnum molle*) with measurements made *in situ* at 24-h intervals. In these determinations, the oxygen-measuring apparatus was specially designed for submerged analyses in the field. Data collected from these experiments were used to make estimates of the potential contribution of *Prochloron* photosynthesis to the ultimate metabolic and respiratory needs of the ascidian. For example, based on certain assumptions concerning translocation magnitude, Olson and Porter (1985) estimated that the phycobionts may contribute between 12 and 31 percent of the carbon respired by the zoobiont. This corresponded to an integrated (24-h) photosynthesis:respiration ratio of 0.62.

To date, the most exhaustive analyses of photosynthesis in ascidian–*Prochloron* phycozoans are found in papers by Alberte et al. (1986, 1987). The first paper (1986) described analyses of the effects of temperature and irradiance (light intensity) on photosynthesis of *Prochloron* cells freshly isolated from intact phycozoans (*Lissoclinum patella*). An interesting aspect of this work was the acclimation experiments. In these experiments, phycozoans collected from the field were separated into two groups. One group was maintained under high light intensity, the other under low light. After several days, photosynthesis measurements made on phycobionts isolated from each group showed pronounced differences. For example, the high-light phycobionts had higher photosynthetic and respiration rates than the corresponding low-light phycobionts. There were also differences in several other photosynthetic features. Apparently the *Prochloron* phytosynthetic systems are capable of light acclimation. In other experiments Alberte et al. (1986) showed that photosynthesis was fairly sensitive to temperature and that at temperatures below 28°C there was a radical decrease in photosynthesis. The authors hypothesize that this stenothermal nature of *Prochloron* photosynthesis may be a prime factor in limiting the distribution of the alga to tropical waters.

In a related series of investigations, Alberte et al. (1987) analyzed photosynthesis of phycobionts *in hospite* and examined intact phycozoans representing six different didemnid species collected from shallow, exposed (high light intensity) and shaded environments. Their data showed that intact phycozoans exhibited photosynthetic acclimation and that photosynthesis (on a daily basis) was sufficient to fuel a significant portion of zoobiont respiration. For example, they estimated that in *L. patella, Prochloron* photosynthesis may contribute 30 to 56 percent of the reduced carbon required by zoobiont respiration. From this work and that cited earlier, it is clear that *Prochloron* cells are capable of significant photosynthesis and, if we make certain assumptions, can provide substantial amounts of reduced carbon to the ascidian. Moreover, the research showed that the *Prochloron* cells exhibit a typical photosynthesis light-intensity response and are capable of photosynthetic rate compensation or photosynthetic acclimation (so-called photoadaptation).

Translocation

Are organic molecules translocated by the phycobionts to the zoobionts, and, if so, what are they?

Radioactive tracer methodology is used to track the flux of organic carbon from phycobiont to zoobiont. Typically, $^{14}CO_2$ as ^{14}C-sodium bicarbonate is added to experimental vessels containing phycozoans or isolated phycobionts. After a period of photosynthesis, the bionts are analyzed for the presence of radioactively labeled molecules. A parallel set of analyses is usually performed on specimens labeled in the dark. In cases where the entire phycozoan is exposed to label, the phycobionts are isolated from the phycozoan after labeling. With most didemnid species, complete and clean physical separation of the bionts is difficult, if not impossible. This means that, after harvesting the phycobionts, the remaining ascidian tissues will be contaminated with residual *Prochloron* cells. Despite this difficulty, it is possible to do labeling studies that provide qualitative and semiquantitative information concerning the amount and nature of organic molecules originating from phycobiont photosynthesis.

Labeling experiments have provided several lines of evidence indicating the translocation of organic molecules from phycobiont to zoobiont. Akazawa et al. (1978) showed that $^{14}CO_2$ is fixed in *Prochloron* by photosynthesis and, in addition to appearing in typical photosynthetic intermediates, may be found among products of the zoobiont's metabolism, presumably formed from precursors originating from the phycobiont. Interpretation of their data must be guarded, however, as their determinations involved analyzing extracts of whole colonies, making it difficult to determine the origin (from ascidian or from *Prochloron*) of some of the radioactive products. Similar experiments were performed by Thinh and Griffiths (1977). Using isolated *Prochloron* cells incubated *in vitro*, Fisher and Trench (1980) showed that the algae could fix $^{14}CO_2$ and liberate some of the labeled organic substances (mainly glycolate) to the surrounding medium, though the actual percentage translocated was low (only 7 percent of the radioactive carbon fixed in the light). This work demonstrated that the algae have the potential to supply some organic materials to the ascidian. Pardy and Lewin (1981), by incubating intact phycozoan colonies (*Lissoclinum patella*) with $^{14}CO_2$ in the light and then analyzing phycobiont cells and zoobiont tissues independently, showed that several specific classes of host molecules (including proteins and nucleic acids) became labeled in due course as a result of the zoobiont's metabolism. The best quantitative analyses of translocation from *Prochloron* cells to the ascidian host are those based on labeling data of Griffiths and Thinh (1983).

What is the impact of translocation products on host nutrition, growth, and survival?

The quantitative studies of Griffiths and Thinh (1983), Olson and Porter (1985), and Alberte et al. (1985–1986) clearly show that the *Prochloron* cells can support a significant fraction of the host's energy demands. Moreover, Pardy and Lewin (1981) have shown that photosynthetically fixed carbon eventually becomes distributed among all the major biochemical constituents (proteins, carbohydrates, lipids, nucleic acids) of the zoobionts. Direct evidence for the impact of translocated substance(s) on the growth of the phycozoan has been provided by Olson (1986). In his experiments, Olson (1986) maintained *Didemnum molle* phycozoans under various light levels, including total darkness. The experimental chambers were maintained at a depth of 2 m and were continuously flushed with fresh seawater. He found that the growth of the phycozoans increased with increasing light with a maximum growth enhancement of up to 40 percent for those specimens maintained in full sunlight. Some of the phycozoans maintained in the dark gained weight, though at a much lower rate than their illuminated counterparts. After nine days in the dark, two specimens had completely lost their complement of *Prochloron* chlorophylls (as judged by the absence of chlorophyll in extracts of the phycozoans). Aside from the reduced growth rate, the apparent lack of phycobionts had no adverse effect on the ascidian.

Are there host factors that control or otherwise affect translocation? Is there evidence for transfer of materials from the host to the algae?

There have been no systematic analyses of factors affecting translocation of organic materials from *Prochloron* cells *in vitro* or *in hospite*, and we do not yet know whether the host can modulate the release of nutrients by the algae, as occurs in some other symbioses. There are no published accounts of studies examining the possible transfer of organic or other molecules from their ascidian host to the *Prochloron* cells, nor on the ability of the *Prochloron*-bearing ascidians to acquire nutrients dissolved in the surrounding water. Certainly these topics will receive greater attention as our efforts to grow *Prochloron* in pure culture continue.

Formation of the Symbiosis

How is the phycozoan established?

At least two forms of the *Didemnum* phycozoan can be recognized: those where the phycobionts live inside the colony and those where the phycobionts live on the ascidian's surface. About 15 species of didemnid exhibit endocolonial symbiosis, with the *Prochloron* cells either embedded in the ascidians' tests or lining their cloacae. These species give rise to larvae with special morphological adaptations for transporting *Prochloron* cells to new settlement sites (Kott, 1977, 1980, 1982), thereby ensuring a continuity of the symbiosis. Each new ascidian colony is endowed with a complement of *Prochloron* cells, which presumably multiply with colony growth. In these species, establishment of the symbiosis by other means has not been reported.

There are more than 20 species of ascidians that may have *Prochloron* cells on their surfaces (Kott, 1984), sometimes associated with cyanophytes. In such species not all specimens are infected (Kott, 1984). We do not know how these extracolonial symbioses are established, though it can be supposed that loose *Prochloron* cells in the plankton could be dispersed and eventually settle on receptive hosts. In fact, Cheng and Lewin (1984) have reported the occurrence of a few free *Prochloron* cells in water overlying *Didemnum* beds.

What is the degree of specificity exhibited by the bionts forming the phycozoan?
Is there evidence for the existence of specific phycobiont–zoobiont recognition mechanisms?

The papers by Lewin (1979), Cheng and Lewin (1984), Kott (1980, 1982, 1984), Müller et al. (1984), Parry (1985) and Cox (1986) emphasize the wide distribution of *Prochloron* among ascidian species and other marine invertebrates. The tendency of this alga to associate with a variety of ascidians could be taken to indicate that the phycobiont is not specific (in symbiosis jargon it could be said that the symbiosis is "non-obligate"). In my mind this begs the question. Observations that the alga can reside on a sea cucumber (Cheng and Lewin, 1984) or in a sponge (Parry, 1985) suggest that these animals have something in common with ascidians. It could be argued that *Prochloron* cells have a high affinity for a factor or *niche* provided by certain didemnid ascidians that is also present in some other invertebrates. The basis of this factor or niche, whether it be chemical, physical, or a combination of both, is unknown.

What features or preadaptations of the zoobionts potentiate the activities or persistence of the phycobionts?

Clearly, didemnid hosts possess certain special features, such as translucency, that permit a degree of photosynthesis by the internal algae. Transport of *Prochloron* cells by larval didemnids is another example of an ascidian adaptation. Olson (1983) described an interesting series of experiments and observations on the settling behavior of larvae from *Didemnum molle*. His data showed that the times of larval release and the settling sites selected appear to be programmed to optimize the photic environment of the phycobionts. Release of larvae occurred around midday in contrast to the dawn–dusk larval release typical of nonsymbiotic ascidians. Larval attachment to the substrate took place on dimly lighted substrates. After metamorphosis the growing colonies, now each consisting of many individual zooids, crept to brighter sites. If larvae were forced to settle in areas of relatively high light flux, the *Prochloron* cells tended to die, and eventually so did the colony. Thus, selection of an appropriately illuminated environment is critical to the survival of the nascent phycozoan.

Not all *Prochloron*-bearing ascidian colonies have the ability to move around by creeping. Some exhibit displacement through a succession of growth and regression cycles (Birkeland et al., 1981). Whether these growth movements have a phototactic basis has not yet been thoroughly studied, though Birkeland et al. (1981) recorded that most colony movement takes place during daylight hours.

It seems that the larval release time and colony motility of some didemnids may affect the health and vitality of the developing phycozoans. Would this behavior be evident in the absence of *Prochloron*, or if algal processes were inhibited? In other words, are the observed release times and motility specialized features of the intact phycozoan or behavior typical of didemnids with or without *Prochloron* symbionts? A partial answer appears in Olson's (1986) study of the effect of light intensity on phycozoan growth, cited earlier. Phycozoans maintained in his dimly lighted or dark chambers exhibited a comparatively high degree of motility, generally an upward migration along the walls to the roofs of the experimental chambers. Although this apparent "light-seeking" response was probably a geotaxis, it was accentuated in the dimmest and dark chambers. One possible conclusion is that the phycozoans were prowling on behalf of the phycobionts for light, which in marine environments tends to be "up."

Finally, Olson (1986) observed a remarkable morphological transformation in the phycozoans maintained in dim light or in darkness. The normal, conical form of the *Didemnum molle* phycozoan was replaced by a flattened morphology, thereby increasing the phycozoan's surface area. In turn, the phycobiont's exposure to light would be increased, thus potentiating photosynthesis. Olson (1986) suggested that this morphological compensation might be a tradeoff, as filter-feeding in the flattened forms might be less efficient.

In a provocative work, Müller et al. (1984) presented evidence that extracts of *Did. molle* had cytostatic activity, to which *Prochloron* cells were apparently immune, but which may suppress or inhibit potentially competing organisms from growing in the ascidian. One difficulty with extrapolating the results of this research to the intact phycozoan is that the bioassay used for measuring cytostatic activity is "reduction in growth" of mouse lymphoma cells. The findings would be relevant if the authors showed that the cytostatic agent also inhibited growth of marine bacteria or algae from the phycozoan's natural environment. Incidentally, it would be interesting to determine whether preparations of mucus from a *Prochloron*-bearing sponge or sea cucumber, or mucus from surface-infected ascidians, show lectin or cytostatic activity.

In the same paper, Müller et al. (1984) reported that mucus from *Did. molle* contains a factor with lectin activity, as indicated by a hemagglutinin test. Isolated *Prochloron* cells suspended in filtered seawater agglutinated, forming clumps, whereas in the presence of didemnid mucus the cells tended to remain dispersed. Furthermore, evidence was presented suggesting that the *Prochloron* cells divided when mucus was present but not in its absence, thus indicating that the proposed lectin might be necessary for promoting growth of *Prochloron* cells *in vitro*. However, their bioassay for the effect of the lectin was the division index of suspensions of *Prochloron* cells incubated with and without mucus for 15 hours. They presented no data on doubling times, specific growth constants, or increases in cell number, wet-packed cell volume, or suspension turbidity. The authors did not adequately demonstrate that the *Prochloron* cells were, in fact, actively growing in their suspensions. The increased division index could have resulted from the carry-over of cells harvested in early stages of division. It would be more convincing to see a suspension culture of *Prochloron* cells stimulated to undergo several rounds of division in response to the presumptive growth promoters in ascidian mucus.

Finally, Griffiths and Thinh (1987) have shown that isolated *Prochloron* cells (from *Lissoclinum patella*) have a higher light-saturated level of photosynthesis than corresponding cells *in hospite*. Moreover, they found that phycobionts *in hospite* incorporated a greater proportion of photosynthetically fixed carbon into protein, whereas isolated cells have a correspondingly greater proportion in the alcohol-soluble fraction. These findings show how some aspects of the phycobionts' photosynthetic and metabolic activities are affected when they are part of (or separated from) a phycozoan.

Biochemical Interactions Between the Symbionts

Is there any evidence for biochemical integration?

Biochemical integration implies that the intact phycozoan has certain metabolic capabilities not possessed separately by the individual bionts. One example of this is the nitrogenase activity exhibited by the *Prochloron–Lissoclinum patella* phycozoan. This activity is apparently present only in the intact phycozoan, not in the isolated bionts (Paerl, 1984). Other *Prochloron*–ascidian phycozoans do not show this effect. Paerl (1984) suggested that this restricted occurrence of nitrogenase activity may reflect particularly low levels of available nitrogen in the habitats where *L. patella* is found. At the very least this seems to be an example of biochemical complementation if not direct integration.

Other Interactions Between the Symbionts

Can the phycozoan be disassembled and reconstituted experimentally?

Experimentally produced viable aposymbionts (algae-free hosts) have been invaluable for the study of other algae–invertebrate symbioses. Aposymbionts are useful as controls for analyzing the effects of the symbiosis on host growth, reproduction, and survival. The experimental resynthesis of phycozoa by the introduction of algal cells into aposymbiotic hosts has been extremely useful in determining the extent of recognition phenomena and in studying processes of symbiont reacquisition in a variety of algal symbioses. Olson's work (1986), cited earlier, points to the maintenance of the phycozoans in the dark as a potential method for eliminating the phycobionts, at least from the *Prochloron–Didemnum molle* phycozoan. Experiments should be done to see if aposymbiotic ascidians survive–thrive–reproduce over an extended period. If this should be the case, many interesting experiments concerning the reestablishment of the phycozoan then become feasible.

Is the ratio of algal biomass to host biomass regulated, and if so, how?

In most phycozoans there are mechanisms operating to optimize and regulate the ratio of host tissue to symbiont cells. In *Prochloron*–ascidian phycozoa, what prevents the algae from overgrowing the host or clogging the cloacal ducts? There are still no ready answers to this question. Although there generally appears to be a stable phycobiont:zoobiont ratio, this has not been substantiated by experiment or quantitative analysis. As to regulatory mechanisms, Griffiths and Thinh (1987) hypothesize that the ascidian, by means of light attenuation perhaps in combination with CO_2 limitation, may act to slow the growth of the phycobionts, thereby suppressing overgrowth.

Clearly, our knowledge of the biology of these fascinating phycozoans has increased greatly during the last 10 years. However, as the responses to the various questions posed in this chapter suggest, much remains to be discovered about these *Prochloron*–ascidian assemblages and how they work.

References (Additional to Bibliography)

Lewin, R. A. Symbiosis and parasitism: definitions and evaluations. BioScience 32:254–256; 1982.

Pardy, R. L. Phycozoans, phycozoology, phycozoologists? *In:* Algal symbiosis: a continuum of interaction strategies, Chap. 1 (L. J. Goff, ed.) Cambridge University Press; 1983:5–17.

Chapter 4

Physiological and Cellular Features of *Prochloron*

Randall S. Alberte

Introduction

Prochloron and its associations with tropical didemnid ascidians have attracted much attention from biologists in the past decade. Though the research ranges from purely ecological studies to molecular aspects of DNA and RNA, much of the effort to date has been directed towards an understanding of the photosynthetic physiology and biochemistry of *Prochloron*. Considerably less attention has been paid to the physiology and metabolic features of the symbiotic host species, and even less to physiological and biomechanical features that ensure the success and define the nature of the symbiosis. An attempt will be made here to address salient features of previous physiological investigations that point to problems for future research. Some of these must underlie the nature of the symbioses between the extracellular symbiont, *Prochloron*, and a variety of host species, including colonial didemnid ascidians, one or two sponges, and at least one holothurian. Lastly, an effort will be made to assess current and future potential contributions of physiological and cell biological studies to an understanding of the phylogenetic affinities of *Prochloron*.

The author would like to acknowledge the enthusiasm and persistence of Ralph Lewin, Lanna Cheng, and Nancy Withers in initiating him into the world of *Prochloron*. His firsthand exposure to *Prochloron* in the nonlyophilized state, in the beautiful surroundings of Palau, West Caroline Islands, served to pique his research interest. During the preparation of this commentary he was partially supported by NSF grant OCE 86–03369.

Photosynthetic Features of *Prochloron*

Carbon Metabolism

Prochloron cells isolated from *Lissoclinum patella* colonies show good activity of ribulose 1,5-bisphosphate carboxylase-oxygenase (RUBISCO), with a K_m similar to that of the enzyme from cyanobacteria (Andrews et al., 1984). In addition, another key enzyme in the Calvin–Benson cycle, phosphoribulokinase, is active in these cells (Berhow and McFadden, 1983). Labeling studies with $H^{14}CO_3^-$ in *Prochloron* cells from *Diplosoma virens* indicate that the three-carbon acid, 3-phosphoglycerate, is the first product of carbon fixation (Akazawa et al., 1978) and that a C-3 or Calvin–Benson pathway for reductive metabolism is the principal means by which *Prochloron* assimilates carbon (Fig. 4.1). CO_2 must be the inorganic carbon substrate for fixation, as RUBISCO uses only CO_2 and not $^-HCO_3$ (see Raven and Beardall, 1981), and therefore *Prochloron* must have an active carbonic anhydrase (CA) to convert $^-HCO_3$, the commonest form of inorganic carbon in seawater, to CO_2 for photosynthetic carbon fixation (see Fig. 4.1). Preliminary evidence (T. D. Sharkey, pers. commun.) indicates that the $\delta^{13}C$ value of *Prochloron* cells from *L. patella* is $-22.59^0/_{00}$, a value typical of C-3 photosynthesis where CO_2 is the substrate for fixation. Although probably supplied directly to *Prochloron* cells *in hospite,* host respiratory CO_2 alone could not support maximum rates of photosynthesis, since colony P:R ratios significantly exceed 1.0 and can be as great as 6 (see the following discussion; Alberte et al., 1987).

Experiments involving $H^{14}CO_3^-$ have yielded useful data to help evaluate metabolic interactions in these symbiotic systems. Most measurements of photosynthate released both from isolated cells and from cells *in hospite* in a number of different ascidian species range from 15 to 50 percent of the total carbon fixed (Kremer et al., 1982; Griffiths and Thinh, 1983, 1987; Alberte et al., 1986), though Fisher and Trench (1980) reported values around 7 percent for *Dip. virens.* The findings indicate that organic compounds released from *Prochloron* can support, in part, the carbon demand of the host. In fact, Griffiths and Thinh (1987) reported that 20 percent of the carbon fixed by *Prochloron in hospite* (*L. patella*) can be recovered in the TCA-insoluble fraction, most of which was in the test material of the host. Pardy and Lewin (1981) found that nearly four times as much labeled material was incorporated by host tissues of *L. patella* and *Dip. virens* in the light as by colonies in darkness. Thus, there is good evidence for the transfer of photosynthetically fixed carbon from *Prochloron* to host tissues, which would fulfill at least one requirement for a "true" symbiosis. The precise nature of the compounds transferred from *Prochloron* to its host *in hospite,* however, is unknown. Some demonstrated or postulated pathways are indicated diagramatically in Figure 4.1.

An important feature of the work by Pardy and Lewin (1981) is that the levels of dark incorporation of ^{14}C by *Prochloron in hospite* are significant (ca. 20 percent of light rates), though dark carbon fixation rates by isolated *Prochloron* cells are insignificant ($\leq$1–2 percent of light rates; Fisher and Trench, 1980; Alberte et al., 1986; Griffiths and Thinh, 1987). β-carboxylation, or the addition of CO_2 to phosphoenolpyruvate (PEP) by PEP carboxykinase (PEP-CK), reported in the cytosol of some algae (Kremer, 1981), may provide a means by which respiratory CO_2 generated by the host during the night could be utilized by *Prochloron* for carbon gain or for transport to its host. If so, it might be expected that the nature of the carbon compound(s) transferred at night (potentially amino acids like aspartate, asparagine, glutamine, etc.) may differ from those transferred during light-dependent photosynthesis where RUBISCO is the carboxylating enzyme.

It has been reported that the chief form of carbon released from *Prochloron* cells *in vitro* is the two-carbon organic acid glycolate (Fisher and Trench, 1980), probably produced as one of the oxidation products of ribulose bisphosphate through the oxygenase activity of RUBISCO (the other product being 3-phosphoglycerate) (see Fig. 4.1). This process,

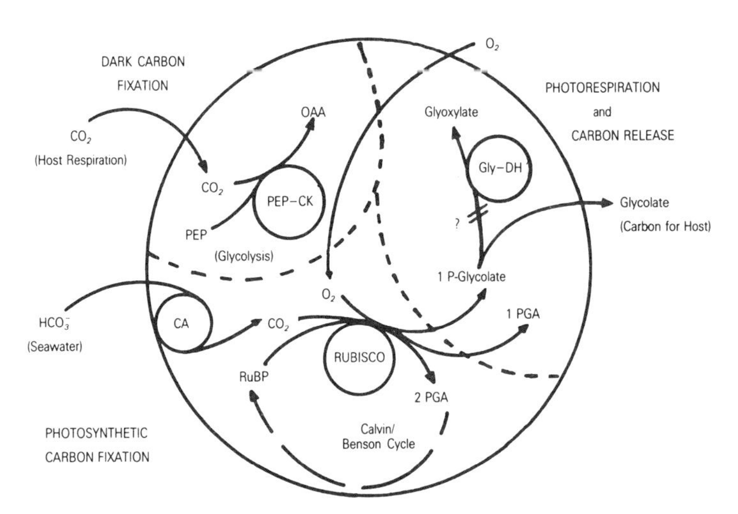

Figure 4.1 Carbon metabolism features of *Prochloron* cells either known or proposed (see text). Abbreviations: CA—carbonic anhydrase; Gly-DH—glycolate dehydrogenase; OAA—oxaloacetic acid; PEP—phosphoenolpyruvate; PEP-CK—phosphoenolpyruvate carboxykinase; PGA—phosphoglyceric acid; RuBP—ribulose bisphosphate; RUBISCO—ribulose 1,5-bisphosphate carboxylase-oxygenase.

termed photorespiration (see Raven and Beardall, 1981, for details), is favored under conditions of high light and high temperature. Therefore, photorespiration may be greatly enhanced during daytime emergence and in exposed sites, where *Prochloron* symbioses are commonly found (Kott, 1980; Ryland et al., 1984; Alberte et al., 1986, 1987). Photorespiration is also favored when either intracellular CO_2 levels are low or O_2 levels are high, which could occur under conditions of saturating rates of photosynthesis typical of *Prochloron* (Alberte et al., 1986). A high activity of carbonic anhydrase (CA), which can serve to concentrate CO_2 inside cells, may depress photorespiratory activity in algae and aquatic angiosperms (Lucas and Berry, 1985) and serve to regulate glycolate production by *Prochloron*.

Photorespiratory production of glycolate (which is light-dependent) reduces the total amount of carbon fixed for growth, because only one (instead of two) phosphoglycerate (PGA) molecules per CO_2 fixed is available to the reductive pentose phosphate pathway of the Calvin–Benson cycle (see Fig. 4.1). This would appear to be an essential if not critical feature of carbon metabolism in *Prochloron* involved in an autotrophic, mutualistic symbiosis. In most algal species studied, glycolate produced through photorespiration is further metabolized, first to glyoxylate and then to glycine and/or serine, with the production of CO_2. In *Prochloron* glycolate is apparently released and not subject to further metabolism. In algae and cyanobacteria (Raven and Beardall, 1981) glycolate is oxidized to glyoxylate by glycolate dehydrogenase, which in cyanobacteria is a thylakoid-bound enzyme (Codd and Sellal, 1978).

It is possible that *Prochloron* and other zoochlorellae that release glycolate lack glycolate dehydrogenase (Gly-DH) activity, thereby preventing further metabolism of glycolate and consequently making significant quantities available for release (see Fig. 4.1). Intracellular glycolate is highly toxic in plants and never accumulates to significant concentrations in the cytosol; instead, if not further metabolized, it is released from the cell. The possibility that glycolate dehydrogenase is present in *Prochloron* but that its activity is regulated by some compound from the host cannot be discounted. If photorespiratory metabolism is high in *Prochloron*, then glycolate metabolism, by reducing the concentration of intracellular oxygen, may promote light-dependent nitrogen fixation, which has been demonstrated in *Prochloron* by Paerl (1984; see the following discussion), and minimize potential toxicity due to supraoptimal concentrations of oxygen. Superoxide dismutase activity in alga–invertebrate symbioses has been implicated as a means of protection from high internal oxygen tensions that develop during high rates of photosynthesis (Dykens and Schick, 1982). The presence and activity of this enzyme in *Prochloron*–ascidian symbioses is unknown.

Photosynthesis–Irradiance Relationships

Rates of photosynthesis in isolated *Prochloron* cells from *Dip. virens* and *L. patella* are as great as or greater than those found in many free-living cyanobacteria and eukaryotic algae (Critchly and Andrews, 1984; Alberte et al., 1986; Griffiths and Thinh, 1987; see Lewin et al., 1985, for summary). *Prochloron* cells isolated from *L. patella* colonies maintained at about 400 μE m^{-2} s^{-1} (one fifth of full sunlight), photon fluxes only slightly higher than those under which the colonies grow normally, show strong photoinhibition of oxygen evolution (Alberte et al., 1986). When the intact colony photosynthesis is examined in *L. patella,* no photoinhibition of photosynthesis is observed even at photon fluxes exceeding those found in the natural habitat (Alberte et al., 1987). Photoinhibition of photosynthesis has not been observed in *Prochloron in hospite* even in well-illuminated colonies of hosts collected from shallow subtidal or intertidal areas (Olson and Porter, 1985; Alberte et al., 1987), except in species that form thin, transparent colonies confined to relatively low-flux environments (e.g., *Dip. similis* and *L. punctatum;* Alberte et al., 1987). Thus, internal self-shading and light attenuation by the host must be important in preventing photoinhibition *in situ*.

Host tissue can attenuate 60 to 80 percent of the light reaching *Prochloron* cells in the cloacal cavities of *L. patella* (Alberte et al., 1986; Griffiths and Thinh, 1987) and *T. cyclops* (Griffiths and Thinh, 1987). Therefore, even though isolated cells may show photoinhibition of photosynthesis at light levels exceeding those that the colonies experience, the attenuation of light by certain host species probably ensures that *Prochloron* cells *in hospite* are never photoinhibited even during episodic exposure (during emergence) to very high quantum fluxes (>2,000 μE m^{-2} s^{-1}). Epizoic *Prochloron* cells found on synaptulid worms in Palau (Cheng and Lewin, 1984) or on colonial ascidians in the Gulf of California (McCourt et al., 1984) are probably not photoinhibited in their natural habitats, even at midday, since the symbioses are normally confined to shaded locations.

The photosynthesis–irradiance (P–I) relationships of *Prochloron* cells isolated from hosts growing in diverse light environments indicate that they are capable of positive photoadaptation. The initial slope of the P–I curve is considerably steeper in cells isolated from colonies growing under low photon-flux conditions than in those from high-flux colonies (Alberte et al., 1986). This feature ensures a high efficiency of light utilization at low quantum fluxes, while lower light levels are required to saturate photosynthesis (I_k) and to achieve photosynthetic compensation (I_c). In addition, the P–I relationships of *Prochloron* cells reflect *in hospite* light regimes and not fluxes incident upon the colonies (Alberte et al.,

1987). As a result, *Prochloron* cells are able to maximize utilization of the daily photosynthetic quantum flux available to them *in hospite* (see the following discussion; Alberte et al., 1986, 1987).

It is clear that *Prochloron* cells, at least in association with *L. patella* from Palau, are phenotypically plastic in their responses to light (Alberte et al., 1986). This feature may be important to their success in hosts that inhabit light environments ranging from fully exposed reef flats to the undersides of coral branches, rocks, or rubble, and epiphytic on roots in dense mangrove stands (see Lewin et al., 1983, 1985). Whether *Prochloron* cells isolated from other hosts are equally plastic remains to be examined. It is possible that there are distinct photosynthetic ecotypes of *Prochloron* in different host species inhabiting such different light environments as *Dip. similis,* found only in low-light environments, and *L. voeltzkowi,* typically found in fully exposed sites. Comparative studies of P–I relationships of six *Prochloron*–ascidian symbioses from a variety of light habitats in Palau do not reveal any compelling evidence for ecoptypic differentiation among the *Prochloron* populations in the different hosts. Rather, the evidence strongly supports the notion that *Prochloron* is phenotypically plastic and can adapt its photosynthetic features to a broad range of light regimes.

Respiratory Behavior and Carbon Balance in *Prochloron*

The relationships between photosynthesis and respiration in *Prochloron* have been examined. The ratios of net photosynthesis to respiration (P:R) exceed 5 in cells obtained from colonies growing in fully exposed sites (Griffiths and Thinh, 1983; Alberte et al., 1986) and can be much greater in cells from colonies growing in shaded environments (e.g., P:R = 16). It has been shown that *Prochloron* respiration is extremely sensitive to environmental photon-flux levels; on a cellular basis, the respiration rates in cells obtained from colonies in exposed sites can be nearly 10 times as high as those from shaded sites (Alberte et al., 1986). Despite differences in light regime between fully exposed colonies and those growing at 20 percent of full sunlight, net carbon gains for *Prochloron* cells from both extremes reach levels of about 70 g C g Chl^{-1} d^{-1} (Alberte et al., 1986), permitting not only rapid growth of the *Prochloron* cells but also transfer of significant amounts of carbon from the symbiont to the host (see the following discussion).

If we assume that 30 percent of the carbon gained by *Prochloron* is released to the host, then, calculated from the preceding data, some 50 g C g Chl^{-1} d^{-1} would still be available for growth and maintenance of the algal symbiont. Published data on two marine phytoplankters, the diatom *Skeletonema costatum* (Gallagher and Alberte, 1985) and a marine cyanobacterium *Synechococcus* sp. (Barlow and Alberte, 1985), indicate that neither the eukaryote nor the autotrophic prokaryote can achieve rates of carbon gain approaching 50 g C g Chl^{-1} d^{-1}. Thus, it appears that the photosynthetic machinery and carbon metabolism of *Prochloron* cells are well suited for existence in a symbiotic association where there may be significant carbon losses to the host (see the following discussion for details).

Properties of the Photosynthetic Pigments and Membranes of *Prochloron*

Pigments and Pigment–Protein Complexes

Unlike typical prokaryotic oxygen-evolving autotrophs, *Prochloron* uses chlorophyll *b* and the light-harvesting chlorophyll *a*/*b* protein (LHC II) rather than chlorophyll *a* and phycobiliproteins for photosynthesis (Lewin and Withers, 1975; Withers et al., 1978). Typically, *Prochloron* from a diversity of hosts exhibits high chlorophyll *a* : *b* ratios, in the range of 4–9 (Lewin et al., 1983; Paerl et al., 1984), and its accessory pigments include carotenoid species characteristic of both eukaryotic algae and cyanobacteria (Withers et al., 1978; Paerl et al., 1984; Foss et al., 1987).

Studies of membrane fractions separated from disrupted *Prochloron* cells revealed that the plasmalemma contains proportionately more carotenoids, particularly zeaxanthin, than the thylakoid membranes (Omata et al., 1985), a feature common to cyanobacteria. Further, the major thylakoid polypeptide, the light-harvesting chlorophyll *a*/*b* protein, shows light-independent phosphorylation, a feature unknown in eukaryotic green algae. However, the photosynthetic lamellae isolated from *Dip. virens Prochloron* have a polypeptide composition similar to that of green algae and higher plants (Schuster et al., 1984).

The pigment–protein composition, apparent molecular weights and spectral features of the chlorophyll–protein complexes from *Dip. virens* symbionts (Schuster et al., 1984, 1985; Hiller and Larkum, 1985) appear similar to those of *Prochloron* from *L. patella* (Withers et al., 1978). However, antibodies to spinach chlorophyll *a*/*b* protein do not cross-react with any thylakoid polypeptides of *Prochloron* from *Dip. virens*, suggesting differences in primary structure between higher plant–chloroplast apoprotein and that of *Prochloron*. More recently, Bullerjahn et al. (1987) showed that polypeptides of the chlorophyll *a*/*b* proteins of *Prochlorothrix* sp., another prochlorophyte, are also not recognized by antibodies against higher plant LHC II. (In neither case was antigenicity to the holoprotein examined.) Collectively, recent investigations (Schuster et al., 1984, 1985; Hiller and Larkum, 1985) confirm the earlier report (Withers et al., 1978) that the composition, spectral features and properties of the photosystem-I reaction center from *Prochloron* from several different host species are indistinguishable from those known for green algae and higher plants. In addition, antibodies generated against the photosystem-I complex from barley recognize this complex in *Prochloron* and in other oxygenic prokaryotes (Alberte, unpublished data).

Why Has *Prochloron* a Chlorophyll *a* + *b* Light-Harvesting System?

It is of interest to ask why *Prochloron* possesses a light-harvesting system based on chlorophyll *a* + *b* and not one based on phycobiliprotein, as is typical of cyanobacteria. Light quality regime and availability of assimilable nitrogen may have been among the selective pressures important in the evolution of such a light-harvesting system in *Prochloron*.

Prochloron–ascidian associations are commonly found in clear, oligotrophic tropical waters. Many are abundant in shallow water and some (particularly those like *L. voeltzkowi* and *Dip. virens*, which are epiphytic, respectively, on seagrass leaves and on mangrove roots) are emergent for a significant portion of the day. In such habitats there is significant availability of orange and red (640–680 nm) wavelengths of light for photosynthesis; these wavelengths are typically unavailable to more deeply submerged aquatic plants and algae because they are selectively absorbed by the water column (Kirk, 1983). In addition, because the waters are clear with little particulate matter, more blue light (400–450 nm), normally scattered in coastal waters or absorbed by Gelbstoff in such waters, is available for photosynthesis. Thus, *Prochloron*–ascidian habitats may exhibit light quality regimes more characteristic of terrestrial situations where chlorophyll *a* + *b* light-harvesting systems prevail.

In *Prochloron*, the excitation spectrum for chlorophyll *a* fluorescence emission shows maxima in the spectral region 400–475 nm (Schuster et al., 1984), corresponding to the *in vivo* Soret absorption bands of chlorophylls *a* and *b*. In contrast, cyanobacteria, even those with phycoerythrin, show little fluorescence excitation in the blue spectral region (400–500 nm) (Alberte, 1989), indicating that they are unable to use blue or blue-green light effectively to drive photosystem II. Furthermore, the fluorescence excitation of chlorophyll *a* + *b* systems in the spectral region 650–675 nm is more than twice that of phycobilin-based systems when normalized to equal chlorophyll *a* contents of the different cell types (Alberte, unpublished). Thus, in the blue and orange-red portions of the spectrum the light-harvesting capacity is lower in phycobilin-based systems than in systems with chlorophylls *a* and *b*. In many of the habitats where ascidian–*Prochloron* associations occur, these wavelengths are not as limiting as in most other marine benthic situations. Thus, growth habitats characterized by shallow clear waters and daily emergence can provide strong selection pressure for a chlorophyll *a* + *b* light-harvesting system in *Prochloron*.

Selective absorption of different light qualities by the host tissues could also provide strong selection pressure on a light-harvesting system. Recent *in vivo* spectral analyses of *Prochloron*-free tissues of *L. patella* have revealed the presence of compounds that absorb strongly in the green to orange portions of the visible spectrum (Fig. 4.2; Alberte, unpublished data). A narrow absorption band in the blue (460–470 nm) may be attributable to a variety of hemelike molecules found in the blood, but the identity of the peak at about 660 nm is unknown. The 660-nm band cannot be attributable or due to chlorophyll or its degradation products, as there is insignificant absorption between 400 and 450 nm, where the Soret bands for any chlorins or porphyrins would be pronounced. The broad green-orange absorption band (550–620 nm) most probably arises from bile pigments in the blood of the host, produced by degradation of heme compounds and perhaps also from pigments present in the test material (see Britton, 1983).

The localization of *Prochloron* within the cloacal cavities of its hosts

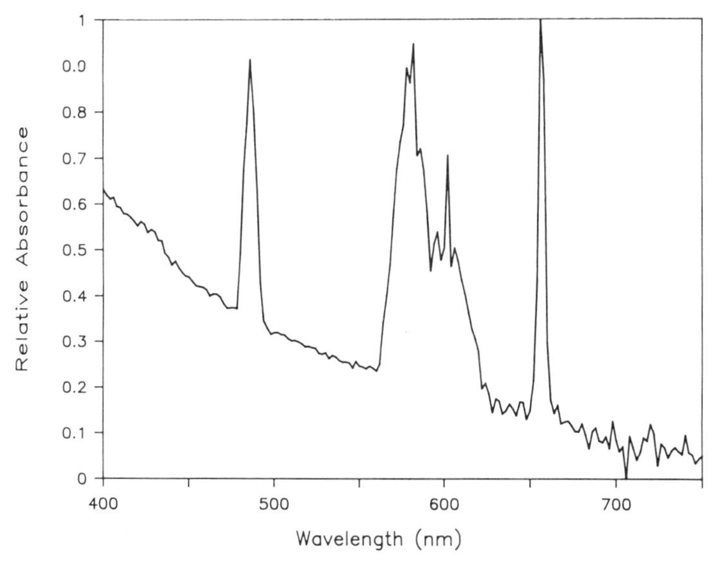

Figure 4.2 Room temperature *in vivo* absorption spectrum of *Prochloron*-free host tissue of *Lissoclinum patella*. Spectrum was obtained on a Hewlitt-Packard Diode Array spectrophotometer (2.0 nm bandpass) and was corrected for scattering (Alberte, unpublished data).

places the cells in a light environment where the host tissues not only attenuate the total quantum flux (functioning like a neutral-density filter), but apparently selectively filter out most of the green to orange wavelengths of light. Consequently, the *in hospite* light environments are poor in wavelengths of light absorbed by the light-harvesting pigment-proteins typical of cyanobacteria, C-phycoerythrin (C-PE), and phycocyanin (PC). Therefore, a phycobiliprotein-based light-harvesting system in *Prochloron* would not permit significant levels of photosynthesis *in hospite,* whereas a chlorophyll *a* + *b* system would.

In this regard, in some of the known associations of cyanobacteria with didemnid ascidians (e.g., *Did. candidum, T. solidum, T. digestum, T. cerebriforme*) (Lewin and Cheng, 1975; Lafargue and Duclaux, 1979; Sybesma et al. 1979; Parry, 1984; Olson, 1986), it has been reported that a *Synechocystis* sp. (coccoid cells) containing an unusual R-type PE is present (Parry, 1984; Parry and Kott, 1988). This urobilin-rich PE is identical to that described and characterized earlier from marine *Synechococcus* spp. (Kursar et al. 1981; Alberte et al., 1984). These PEs, unlike C-PE, have a strong absorption maximum around 490 nm because of the presence of urobilin chromophores which are absent from C-PE. Further, these R-type PEs have their major green absorption bands (530–580 nm) shifted to the blue compared with those of C-PE (Alberte et al. 1984). Therefore, the *in vivo* absorption properties of the R-type PEs found in the symbiotic cyanobacteria in ascidians (Parry, 1984; Parry and Kott, 1988) correspond quite closely to the "green window" created *in hospite* by the selective absorption properties of the host tissues. Thus, the *in hospite* light environment for symbionts in didemnid ascidians is such that either a chlorophyll *a* + *b* system or a phycobiliprotein system based upon a fairly unique R-type PE would be the most suitable system for effective photosynthesis.

Nitrogen availability could also have been an important selective factor leading to the evolution of a chlorophyll *a* + *b* light-harvesting system in *Prochloron.* Based on known features of the photosynthetic unit (PSU) of *Prochloron* (Withers et al., 1978; Alberte et al., 1986), one can estimate that the light-harvesting chlorophyll *a*/*b* protein accounts for about 40 percent of the total chlorophyll, and the PSU has about 250 molecules of the chlorophyll *a*/*b* complex per reaction center (P700 or P680; see Alberte et al., 1986). If we assume that the light-harvesting chlorophyll *a*/*b* protein accounts for all the chlorophyll *b* present and that it has equal proportions of chlorophylls *a* and *b,* then there are 50 chlorophyll *a* and 50 chlorophyll *b* molecules in each light-harvesting chlorophyll *a*/*b* protein, along with 150 additional molecules of chlorophyll *a,* in each PSU (for details see Withers et al., 1978).

Since each light-harvesting chlorophyll *a*/*b* apoprotein (26 kD) binds three chlorophyll *a* and three chlorophyll *b* molecules, it would require about 433 kD of protein to bind the chlorophylls (100 chromophores) in all the light-harvesting chlorophyll *a*/*b* proteins in each PSU. The average molecular weight of an amino acid is about 0.1 kD, and in each PSU there are about 4,330 amino acid molecules in each light-harvesting chlorophyll *a*/*b* protein. Therefore, each PSU would require at least 4,330 atoms of nitrogen for the apoprotein and an additional 400 nitrogen atoms for the chlorin molecules (4 *N* per tetrapyrrole). This would necessitate a minimum of about 5,750 atoms of nitrogen to build the bulk light-harvesting system for each PSU.

One can run through a similar set of calculations for a phycobilin-based light-harvesting system like that of *Anacystis nidulans*, in which each light-harvesting system contains about 150 chlorophyll *a* molecules and 100 bilin chromophores in the phycobilisome (Kursar and Alberte, 1983). Phycocyanin (PC) would contribute about 85 chromophores, whereas allophycocyanin (APC) would contribute about 15 bilins, to yield a bulk light-harvesting system similar to that described above with a PSU size of 250 (for details see Kursar and Alberte, 1983; Kursar et al., 1983). For each 35 kD of APC there are two bilins and for each 37 kD of PC there are three bilins. Thus, 1,048 kD of protein would be required to bind 85 PC bilins, and 280 kD of protein would be needed for the 15 APC bilins. A bulk light-harvesting system of 100 chromophores would contain about 13,280 amino acid residues, or 1,328 kD of protein, containing 13,280 atoms of nitrogen. Adding 400 nitrogen atoms for the bilin chromophores (4 *N* per tetrapyrrole) would need 13,680 atoms of nitrogen. Lastly, because phycobilisomes are also composed of structural or "linker" proteins (accounting for about 15 percent of the total protein in the phycobilisome), an additional 2,000 atoms of nitrogen are required. Thus, to build a conventional cyanobacterial light-harvesting system, or phycobilisome, one would need at least 15,700 atoms of nitrogen.

These calculations demonstrate that it takes about three times as much nitrogen or protein to construct a bulk light-harvesting system of phycobiliproteins as that needed to construct an equivalent PSU (with 250 chromophores per reaction center) using the chlorophyll *a*/*b* protein. Nitrogen limitation could therefore have been important in the selective evolution of the *Prochloron* light-harvesting system.

The chlorophyll *a*/*b* protein (LHC II) in eukaryotes is encoded in a set (five to 10 copies) of nearly identical nuclear genes (for details see Cashmore, 1986). The polyadenylated mRNAs for the proteins are translated on cytoplasmic ribosomes (80s) as precursors, which are later modified on import into the chloroplast by cleavage of a 4-kD leader sequence. It would be of interest to know whether there are, similarly, multiple gene copies in *Prochloron* and whether the light-harvesting chlorophyll *a/b* proteins are synthesized as precursors or are made at the native molecular sizes, as is more typical in prokaryotes where transcription is tightly coupled to translation. Of course, the ability to make chlorophyll *b* chromophores is equally important; the enzymes necessary must be present in *Prochloron*. The methodology is now available to examine these molecular genetic questions as soon as *Prochloron* can be cultured in the laboratory or good-quality DNA can be isolated in the field.

Are any of the genes for the synthesis of phycobiliproteins and their linker proteins present but repressed in *Prochloron?* Recent work (Bryant et al., 1986) has demonstrated that APC and PC genes in cyanobacteria are transcribed on polycistronic messages and translated in their native molecular sizes. The polycistronic messages also encode the linker proteins. DNA probes now available for APC and PC and for the linker proteins would permit one to examine whether these genes are present in the genome of *Prochloron*. Though *Prochloron* need not be grown in culture for such investigations, it is essential that full-length DNA be isolated from the cells and examined by using nucleic acid-blotting techniques with suitable DNA or RNA probes. It is hoped that such studies will be pursued in the near future.

Can *Prochloron* synthesize the necessary chromophores to make either a phycobilisome or a chlorophyll-based light-harvesting system? Although the absence of the ability to make a specific chromophore usually leads to inhibition of synthesis of the apoprotein to which it binds, the absence of a given pigment-protein does not necessarily indicate the inability to synthesize that apoprotein. Bednarik and Hoober (1985) showed that in *Chlamydomonas* the synthesis of chlorophyll *b* probably diverges from the known pathway for chlorophyll *a* synthesis at the formation of a chlorophyll *b* precursor, chlorophyllide *b*, from protochlorophyllide, which is an intermediate common to both pathways. The formation of chlorophyllide *b* is dependent on a reduction step (light-dependent in angiosperms but not in most green algae) that is also required for the formation of chlorophyllide *a*, followed by the oxidation of the methyl group on ring B to an aldehyde group, probably mediated by a specific oxidase common to green algae and higher plants. These steps presumably occur in *Prochloron* but require examination.

Since the phycobilin chromophores PC and APC share the same biosynthetic pathway as chlorophyll *a* through the cyclization of the pyrrole groups, one unique requirement for bilin synthesis is for a heme oxidase, an enzyme that cleaves the cyclized tetrapyrrole precursor to an open-chain tetrapyrrole typical of bile pigments (Troxler et al., 1979). In *Prochloron* the absence of such a heme oxidase might be expected, though it has not yet been sought.

The phylogenetic affinities of *Prochloron* are still unclear. It is anticipated that molecular genetic investigations aimed at delineating genetic features of its photosynthetic apparatus may provide important clues to its origin and its relatedness to cyanobacteria and green algal and plant chloroplasts that have not been revealed by other efforts to date (see Chap. 6).

Physiology of the *Prochloron*–Ascidian Associations

Photosynthesis and Respiration of Symbiotic Didemnids

Unfortunately, little is known about the physiological and feeding characteristics of the ascidian hosts of *Prochloron*. The production of fecal pellets (see Lewin et al., 1983) indicates that they ingest and excrete particulate organic matter, but we know very little about the extent to which feeding is important to their nutrition and the nature of the ingested material. Olson (1986) showed that light dramatically enhanced the growth (wet wt.) of *Did. molle;* however, colonies maintained in darkness did not show weight losses, though they did lose more than 80 percent of their chlorophyll content. Loss of chlorophyll from colonies of *L. patella* held in prolonged darkness (five days) and their subsequent death were observed by Cheng (unpublished observations). Recently, it was observed that colonies of *L. patella* maintained under natural illumination in ultrafiltered seawater (0.2 μm) lost both wet and dry weight over a five-day period, whereas colonies kept under similar conditions but provided with particulate organic matter (phytoplankton) showed net weight gains (Alberte, Zimmerman, Cheng, and Lewin, in prep.). Further, it was shown that when colonies were provided ^{15}N-labeled phytoplankton, significant ^{15}N-enrichment was observed not only in the host tissues but in the symbiont (Alberte et al., in prep.). In a parallel experiment, it was found that significant ^{15}N-enrichment occurred in the host when inorganic ^{15}N was provided for assimilation by the symbiont. Therefore, there appear to be exchanges of nitrogenous compounds between both partners in the symbiosis.

Studies by Pardy (1984) showed that two symbiotic ascidians (*L. patella, Did. ternatenum* = *Did. molle*) have respiratory rates (measured as oxygen consumed in darkness per unit dry weight of animal) greater than those determined for nonsymbiotic ascidians (Goodbody, 1974). This could be attributable to additional oxygen uptake by the *Prochloron* cells (see Alberte et al., 1986). Under all the light levels examined between 225 and 2,000 $\mu E\ m^{-2}\ s^{-1}$, Pardy (1984) found net oxygen evolution from symbiotic colonies, with P:R ratios of about 2.3. For colonies of *Dip.*

virens higher P:R values, around 8.5, were measured by Thinh and Griffiths (1977), whereas Olson and Porter (1985) reported P:R ratios of *Did. molle,* determined *in situ,* as low as 0.62. In the latter studies, dark respiration values were found to be more than twice as high as those reported for *Dip. virens* or *L. patella* by Pardy (1984).

Measured P:R ratios of the algal symbiont alone from *L. patella* were generally about 5 but exceeded 15 under some growth conditions (Alberte et al., 1986). Using data for photosynthesis per gram of chlorophyll of *Prochloron* from *L. patella* and the chlorophyll content per gram dry weight of *L. patella* (Pardy, 1984; dry weight/wet weight ratio = 0.07), one can calculate that a large, well-illuminated colony of about 70 g wet weight, containing about 18 billion *Prochloron* cells with 5.5×10^{-9} mg Chl $cell^{-1}$, would produce approximately 45 μmol O_2 min^{-1}. Therefore, on a typical sunny day with 10 h of light-saturated photosynthesis ($P_{max} = 3 \times 10^{-9}$ μmol O_2 $cell^{-1}$ min^{-1}; Alberte et al., 1986), each algal cell could produce about 0.2 nmol O_2 d^{-1}. The whole colony would have a net production rate of about 20.5 mol O_2 d^{-1}, assuming a loss of about 15.5 mol O_2 d^{-1} consumed by host respiration (based on a *L. patella* colony P:R ratio of 2.3; Pardy, 1984). This value converted to carbon units per chlorophyll would yield a production rate of about 7.5 μg C Chl^{-1} d^{-1}. An aerial production rate for *L. patella* of 13.1 μg C cm^{-1} d^{-1} was determined by Alberte et al. (1987), taking into account net photosynthetic production and respiratory losses due to host and symbiont. Since colonies of *Dip. virens* and *L. patella* show significant daily O_2 production rates, these organisms could make important contributions to the overall benthic productivity in areas where they are found (see Lewin et al., 1983; 1985). However, the P:R features of *Did. molle* indicate only a minor contribution of the symbionts to colony production (Olson and Porter, 1985; Olson, 1986; see the following discussion).

Alberte et al. (1987) found that six *Prochloron*–ascidian associations from Palau have P:R ratios (based on P_{max} and maximal dark-respiration rates of colonies) between 1.6 and 5.9. Five of these associations (*L. patella, L. punctatum, T. cyclops, Dip. virens,* and *Dip. similis*) exhibited P:R ratios greater than 1.0, when the ratios were determined on a daily basis. Dark respiration was assumed to be constant over a 24-h period, and photosynthetic activity was based on the daily periods of light-compensated photosynthesis for the colonies. When colonies of *L. patella* were maintained under high and low photon fluxes, the daily P:R ratios were found to be greater for the low-light colonies than for the high-light ones, primarily because colony respiration was reduced almost 50 percent under low light whereas photosynthesis showed much smaller reductions. Olson and Porter (1985) reported a daily P:R ratio for *Did. molle* of 0.62, which was determined using an *in situ* respirometer. Despite this low P:R ratio, they also reported a 40 percent enhancement in growth of the same populations of *Did. molle* under the ambient light conditions found in a lagoon in Lizard Island, Australia, compared with growth in total darkness (Olson, 1986).

To assess reliably the daily contribution of photosynthetically fixed carbon by the symbionts of *L. patella* to the respiratory demand of their host, Alberte et al. (1987) determined the net carbon production by the symbionts, the carbon losses to respiration by both the symbionts and the host, and used the value of 20 percent for the amount of carbon fixed by the symbionts and made available to the host (see Alberte et al., 1986; Griffiths and Thinh, 1987). Using the formulations of Muscatine et al. (1981), the CZAR (daily symbiont carbon used to support host respiration) was determined for high- and low-light grown *L. patella* colonies. Values of 30 to 56 percent were obtained, indicating that a large and significant portion of the carbon required by host respiration is met by *Prochloron* photosynthesis. *Prochloron*–ascidian CZAR values are in the range known for other symbiotic systems (e.g., Muscatine et al., 1981). By comparing the determined daily P:R ratios for other *Prochloron* symbioses to those of *L. patella* (see earlier), it appears that *Prochloron* makes a significant and probably obligatory contribution to the organic nutrition of most of its hosts (cf., Alberte et al., 1987).

Nitrogen Assimilation

Parry (1985) investigated the metabolism of ^{15}N-ammonium and nitrate by intact colonies of *L. patella*, *T. cyclops*, and *Dip. virens*. After a 2-h labeling period under illumination in seawater containing ^{15}N-ammonium (100 μM), ^{15}N-labeled glutamine was found in the *Prochloron* isolated from all three hosts, but no labeled intermediates were detected in the host tissues. When colonies were incubated with ^{15}N nitrate, in either light or dark, no label was detected in host or symbiont. In the ammonium feeding experiments no labeled glutamine was recovered in the bathing medium, indicating that all the end-products of ammonium assimilation by *Prochloron* were conserved during the incubation period.

The ^{15}N–NMR spectrum of the labeled glutamine obtained in these experiments led Parry (1985) to conclude that the ammonium was assimilated through the glutamine synthetase + glutamate synthetase (GS + GOGAT) pathways, which are highly active in many prokaryotic and eukaryotic photoautotrophs. The apparent lack of nitrate assimilation by *Prochloron in hospite* might be attributable to the fact that nitrate reductase is inducible by nitrate, an ion virtually absent from the habitat, and the incubation periods may have been too short for enzyme induction (Parry, 1985).

Examinations of glutamine synthetase (GS), an enzyme central to ammonium assimilation, in *Prochloron* cells isolated from *L. patella* revealed high activities of this enzyme (Alberte, Zimmerman, Cheng, and Lewin, in prep.). When colonies or isolated *Prochloron* cells were incubated for up to 24 h with 20 μM ammonium, no enhancement of GS activity was observed. These findings indicate that the GS activity is constitutive in *Prochloron* and that the glutamine production in the intact colonies described by Parry (1985) is probably attributable to the activity of this enzyme. In comparison, when intact colonies or isolated *Prochloron* cells of *L. patella* were incubated with 100 μM nitrate for 24 h, high activity of nitrate reductase (NR) was observed (Alberte et al., in prep.). Neither colonies nor freshly isolated cells from colonies collected in the field showed significant NR activity, though induction of NR activity was rapid upon exposure to nitrate. After 4 h of exposure to nitrate, NR activities increased more than tenfold; activity was readily detectable after only 2 h. Thus, it appears that NR is not a constitutive enzyme in *Prochloron*, but shows very rapid induction kinetics.

The data of Parry (1985) and the recent findings of Alberte et al. (in prep.) indicate that *Prochloron in hospite* can effectively use ammonium obtained either as a major nitrogenous waste product of its host (see Goodbody, 1974) or from the water column. In addition, nitrate, when available, can be assimilated. The nitrogen economy of *Prochloron* apparently could be supplemented in some host species by light-mediated dinitrogen fixation (measurable by acetylene reduction), as shown in *L. patella* colonies by Paerl (1984). Nitrogenase, however, is strongly inhibited by ammonium ions as well as by oxygen (Burns and Hardy, 1975), and Paerl was unable to detect nitrogenase activity in *Prochloron* cells isolated from the host, *L. patella*, or in intact colonies of *L. voeltzkowi*, *Did. molle*, on *Dip. virens*. Nevertheless, we should recall that several species of unicellular cyanobacteria are known to fix nitrogen and carry out oxygenic photosynthesis at the same time (Fogg, 1974). Verification of the nitrogen-fixation capabilities of *Prochloron*, by using immunological techniques, is required.

Although there are inconsistencies between the Parry (1985) and Paerl (1984) studies, the activity of the GS–GOGAT system is not constrained by the source of ammonium. That is, ammonium derived from a source external to the cell can be assimilated by the GS–GOGAT system as well as ammonium derived from nitrogen fixation. The real discrepancy lies in Parry's contention that *Prochloron* cells are normally bathed in a medium containing ammonium ions. This assumption is based on determinations of the rates of ammonium release in nonsymbiotic, solitary ascidians, which are strictly dependent on feeding for their nutrition (Goodbody, 1957). Whether symbiotic, colonial ascidians have significant rates of ammonium release, and the extent to which this may be available for uptake by *Prochloron* cells, remain to be examined. The very recent findings of Alberte et al. (in prep.) support the notion that *Prochloron* cells can effectively use ammonium. However, if the symbiont is really "bathed" in ammonium derived from the host, one would not expect to see induction of NR activity when nitrate is supplied to intact colonies, since ammonium is an effective inhibitor of NR activity. Thus, ammonium levels *in hospite* may be so low that they do not inhibit nitrogenase or NR activity.

In this regard, it has been found that *Prochloron* cells *in hospite* take up and assimilate ^{15}N-ammonium, leading to significant percent ^{15}N enrichments in the symbiont cells (Alberte et al., in prep.). In a similar manner ^{15}N-nitrate is readily assimilated by *Prochloron* in intact colonies. The high degree of inorganic nitrogen assimilation by *Prochloron in hospite* argues that the symbionts are not "bathed" in ammonium derived from the host and that, in fact, the symbiont cells appear to be nitrogen limited.

The findings of Paerl (1984) that colonies of other symbiotic didemnids, epiphytic on seagrass leaves and mangrove roots or on hard substrata within a few meters of land in Palau, showed no N_2 fixation may be because they had other sources of ammonium or nitrate, either from terrigenous run-off or from nitrogenous compounds leached from the living substratum. The ability of *Prochloron* cells to fix nitrogen only *in hospite* can be most simply explained by the fact that the host tissues reduce oxygen tensions. This notion is supported by observations that respiration rates of symbiotic didemnids tend to increase when symbiont photosynthesis increases (Alberte et al., 1987), which could keep oxygen tensions low near the symbionts in the cloacae.

Prochloron cells themselves may have a variety of mechanisms for keeping intracellular O_2 tensions low. These include (1) unusually high rates of respiration for a cyanobacterium-like cell (Alberte et al., 1986); (2) high photorespiratory consumption of O_2, which ensures a ready supply of glycolate for release; and (3) considerable cell surface area in close association with animal tissue, favoring rapid diffusion of O_2 out of the cells to support host respiration. Moreover, the symbiotic associations are almost exclusively found in warm tropical waters, where oxygen solubility is lower than in temperate waters. During daylight hours much of the oxygen to support host respiration could be derived from symbiont photosynthesis. Clearly, further work is required to elucidate the metabolic exchanges between host and symbiont. These investigations will require considerations of the pumping features of the host and the fluid dynamics in and around colonies, as well as the partitioning of different metabolites between the partners in the symbiosis.

Obligate Nature of Symbiosis

Prochloron cells are for the most part extracellular, which raises some interesting questions concerning the nature of their symbiosis. Bachmann et al. (1985) found that such cells were absent from colonies of *Did. molle* collected below 40 m in the Maldives (Indian Ocean), although colonies at the same site in shallower water contained the algal symbionts. In the absence of *Prochloron* cells, there appeared to be no major impairment of ascidian growth, though the colonies were only one fifth or one sixth as large. These findings have been supported in experiments by Olson (1986), who found that growth of *Did. molle* colonies from Lizard Island (Australia) increased with increasing quantum flux, and colonies that had been kept in darkness and had lost their *Prochloron* cells were able to maintain a constant colony weight. Similar types of experiments conducted on *L. patella* revealed that, after colonies had been kept in darkness for five to 10 days, loss of *Prochloron* cells accompanied death of the animal host (Cheng, unpublished findings). It is not unreasonable to suggest that the obligate nature of the symbiosis between *Prochloron* and certain hosts may depend on heretofore unexamined features of the host or the host's environment. For example, the symbiotic association of *Prochloron* with certain ascidian species may depend on the availability and abundance in the water column of suitable food material for the animal, or on specific biomechanical features of pumping and feeding in the host species. Likewise, the success or distribution of the host may be dependent on physiological features of, and/or substances (e.g., vitamins) liberated from, *Prochloron*.

The temperature sensitivity of *Prochloron* photosynthesis may be a key factor regulating the global distribution of *L. patella*. It has been found that photosynthesis of *Prochloron* is extremely sensitive to chilling and that cells show weak recovery from even a brief (20-min) exposure to temperatures only 5°C below their ambient growing temperature (Alberte et al., 1986). In fact, the Q_{10} for *Prochloron* photosynthesis between 5° and 30°C is nearly twice as high as that above 30°C. Isolated *Prochloron* cell respiration shows Q_{10} values about half those of photosynthesis over the range 5° to 45°C (Alberte et al., 1986). The Q_{10} value for respiration of intact colonies of *L. patella* is 2.0 over the temperature range 15° to 45°C, whereas below 30°C the values for photosynthesis are greater than 3.0 (Alberte et al., 1987). Since, to date, *L. patella* symbiotic with *Prochloron* has been found only in waters in which ambient temperatures rarely drop below 22°C, and since it appears that *Prochloron* is incapable of carrying out photosynthesis at temperatures below 20°C, though host respiration is little affected by these temperatures, it is likely that the photosynthetic physiology of the algal symbiont determines the thermal distribution of its host and that the association is obligatory.

In this regard, the carotenoid composition and/or content of *Prochloron* also may provide a clue to the apparent inability of this alga to survive in the free-living state (Paerl et al., 1983; Foss et al., 1986). The role of β-β-carotene and other carotenoids in photoprotection is well documented (see Britton, 1983). Certain cyanobacteria (e.g., *Microcystis aeruginosa*) capable of increasing their carotenoid content (specifically, of β-β-carotene) not only can withstand high flux (>1,800 $\mu E\ m^{-2}\ s^{-1}$) of visible and UV irradiation but also show enhanced photosynthetic performance under these conditions (Paerl, 1984; Paerl et al., 1983, 1985). Species able to withstand high UV irradiation show concomitant increases in β-β-carotene, whereas those sensitive to UV (e.g., *Anabaena oscillarioides*) do not. Thus, it is possible that *Prochloron* cells may be unable to adjust their carotenoid content or composition to afford photoprotection under the high photon fluxes of visible or UV light common to the habitats where *Prochloron*–ascidian symbioses are found. In fact, when removed from their hosts, *Prochloron* cells show strong photoinhibition at natural photon-flux levels (Alberte et al., 1986). Certainly in their natural situations the cells are, to some extent, protected from high irradiance and UV damage by the "shading" effects and selective absorption properties of the host tissue (Alberte et al., 1986, 1987).

Although *Prochloron* may not be required to maintain survival and growth of all its animal hosts, we still have no evidence for the existence of free-living *Prochloron*. The association of *Prochloron* with certain ascidians seems to be essential to the success of the alga, but not necessarily to that of the host.

Some Suggested Physiological Requirements for the Culture of *Prochloron*

A perplexing problem, which clearly hinders much of the progress on the biology of *Prochloron*, is that this alga has not been brought into axenic culture, despite considerable effort to date (see Lewin, 1985). A single report indicating that *Prochloron* was a tryptophan auxotroph and that the host (*Dip. virens*) provided this amino acid awaits confirmation (Patterson and Withers, 1982).

We may ask what physiological requirements of *Prochloron* have not been met in culture attempts so far. First, it may be that, when such cells are removed from their hosts, their carbonic anhydrase activity is lost or declines significantly (perhaps due to turnover) and becomes insufficient to meet the CO_2 requirements for growth. *In hospite* a portion of the CO_2 requirement is met directly from host respiration. Attempts should be made to culture *Prochloron* at a lowered pH with relatively high CO_2 concentrations in well-agitated suspensions to enhance direct CO_2 utilization.

Second, *Prochloron* cells *in vitro* may be sensitive to atmospheric levels of O_2 and may live *in hospite* only under reduced oxygen tensions created by the rapid uptake of O_2 by host respiration. For about 12 h each day, at night, *Prochloron* is subjected to reduced oxygen tensions because of the high O_2 demand by the host as well as by itself. It may be possible to culture *Prochloron* cells at low photon fluxes in soft agar, using "stab" culture techniques, to reduce oxygen tensions around the cells. It might also be worthwhile to try to culture cells under a continuous stream of a gas mixture (e.g., 600 ppm CO_2 in dry N_2 or Ar) during illumination periods, to drive off O_2 produced in photosynthesis.

Third, it may be important for the cells to be associated or attached to a matrix or solid substratum similar to that in the colonies. Since the cells divide by binary fission (Lewin et al., 1985), complete separation of the two daughter cells may require a physical substratum that could be provided by a matrix or surface. Glass test tubes may not provide such a suitable surface. The use of glass or polystyrene beads, Sephadex resins (e.g., G-10 or G-25), or even nonosmotically active materials such as Percoll (colloidal silica), might prove successful in suspension cultures.

Fourth, in view of the recent findings of Lewin and Cheng (in prep.) that *Prochloron* cells have an osmotic potential about 20 percent greater than that of seawater and that the cytoplasmic sap contains unusually high levels of serine, it may be possible to use this information and other data on the composition and physical features of the cytoplasm of *Prochloron* to devise more suitable media for the culture of this prochlorophyte.

Conclusion

The past decade has been a highly productive period of investigation of the green prokaryote, *Prochloron* sp. Despite advancements in our knowledge, however, much remains to be done that is critical to our understanding of the nature of this symbiosis between *Prochloron* and its hosts. Not all of this needed work necessitates successful culture of *Prochloron* or of the symbiotic associations, though of course the achievement of such cultures would greatly facilitate research progress. (Investigators with ready access to *Prochloron*–ascidian associations, especially in areas like Palau, Micronesia, however, should be cautioned not to exploit these resources to the point where their abundances or distributions are seriously disturbed or depleted.) The next decade of physiological and cell biological research on *Prochloron* and its hosts holds many promises for advancing our understanding of symbiosis in general.

References (Additional to Bibliography)

Alberte R. S. Optical properties of marine macrophytes and phytoplankton. *In:* Photosynthesis in the Sea (R. S. Alberte, R. T. Barber, and J. A. Connors, eds.), New York: Oxford University Press; 1989 (in press).

Alberte, R. S., A. M. Wood, and T. A. Kursar. Novel phycoerythrins in marine *Synechococcus* spp.: Ecological and evolutionary implications. Plant Physiol. 75:732–739; 1984.

Barlow, R. G., and R. S. Alberte. Photosynthetic characteristics of phycoerythrin-containing marine *Synechococcus* spp. I. Responses to growth photon flux density. Mar. Biol. 86:63–74; 1985.

Bednarik, D. P., and J. K. Hoober. Synthesis of chlorophyllide *b* from protochlorophyllide in *Chlamydomonas reinhardtii y-1.* Science 230:450–453; 1985.

Britton, G. The biochemistry of natural pigments. Cambridge University Press, 1983: 366 pp.

Bullerjahn, G. S., H. C. P. Matthijs, L. R. Mur, and L. Sherman. Chlorophyll–protein composition of the thylakoid membranes from *Prochlorothrix hollandica,* a prokaryote containing chlorophyll *b.* Eur. J. Biochem.; 1987.

Bryant, D. A., et al. The cyanobacterial photosynthetic apparatus: a molecular genetic analysis. *In:* Current Communications in Molecular Biology. Cold Spring Harbor Laboratory Publ., Cold Spring Harbor, N.Y.; 1986.

Burns, R. C., and R. W. F. Hardy. Nitrogen fixation in bacteria and higher plants. *In:* Molecular biology, biochemistry and biophysics, Vol. 21 (A. Kleinzeller, G. F. Springer, and H. W. Wittmann, eds.). Berlin: Springer-Verlag; 1975.

Cashmore, A. R. Nuclear genes in photosynthesis. *In:* Proc. VII Int. Congr. Photosyn. Res.—Providence, R.I. (J. Biggins, ed.), New York: Alan Liss; 1986.

Codd, G. A., and A. J. J. Sellal. Glycolate oxidation by thylakoids of the cyanobacteria *Anabaena cylindrica, Nostoc muscorum* and *Chlorogloea fritschii.* Planta 139:177–182; 1978.

Dykens, J. A. and J. M. Schick. Oxygen production by endosymbiotic algae controls superoxide dismutase activity in their animal host. Nature 297:579–580; 1982.

Fogg, G. E. Nitrogen fixation. *In:* Algal physiology and biochemistry (W. D. P. Stewart, ed.), Botanical Monographs, Vol. 10. Berkeley: Univ. California Press; 1974:560–582.

Gallagher, J. C., and R. S. Alberte. Photosynthetic and cellular photoadaptive characteristics of three ecotypes of the marine diatom, *Skeletonema costatum* (Grev.) Cleve. J. Exp. Mar. Biol. Ecol. 94:233–250; 1985.

Goodbody, I. Nitrogen excretion in Ascidiacea 1. Excretion of ammonia and total non-protein nitrogen. J. Exp. Biol. 34:297–305; 1957.

Goodbody, I. The physiology of ascidians. Adv. Mar. Biol. 12:1–149; 1974.

Kirk, J. T. O. Light and photosynthesis in aquatic environments. Cambridge: Cambridge University Press; 1983.

Kursar, T. A., and R. S. Alberte. Photosynthetic unit organization in red algae: Relationships between light-harvesting pigments and reaction centers. Plant Physiol. 72:409–414; 1983.

Kursar, T. A., H. Swift, and R. S. Alberte. Morphology and light-harvesting complexes from a novel cyanobacterium: Implications for phycobiliprotein evolution. Proc. Natl. Acad. Sci. (USA) 78:6888–6872; 1981.

Kursar, T. A., J. van der Meer, and R. S. Alberte. Light-harvesting system of the red alga, *Gracilaria tikvahiae*. II. Phycobilisome characteristics of pigment mutants. Plant Physiol. 73:361–369; 1983.

Lucas, W. J., and J. A. Berry. Inorganic carbon transport in aquatic photosynthetic organisms. Physiologia Plantarum 65:539–543; 1985.

Muscatine, L., L. McCloskey, and R. E. Marian. Estimating the daily contribution of carbon from zooxanthellae to coral animal respiration. Limnol. Oceangr. 26:601–606; 1981.

Paerl, H. W. Cyanobacterial carotenoids: their roles in maintaining optimal photosynthetic production among aquatic bloom-forming genera. Oecologia 61:143–149; 1984.

Paerl, H. W., P. T. Bland, N. D. Bowles, and M. E. Haibach. Adaptation to high-intensity, low wavelength light among surface blooms of the cyanobacterium *Microcystis aeruginosa*. Appl. Environ. Microbiol. 49:1046–1052; 1985.

Paerl, H. W., W. J. Tucker, and P. T. Bland. Carotenoid enhancement and its role in maintaining blue-green algae (*Microcystis aeruginosa*) surface blooms. Limnol. Oceanogr. 28:847–857; 1983.

Parry, D. L. Cyanophytes with R-phycoerythrins in association with seven species of ascidians from the Great Barrier Reef. Phycologia. 23:503–505; 1984.

Raven, J. A., and J. Beardall. Respiration and photorespiration. *In:* Physiological bases of phytoplankton ecology (T. Platt, ed.). Can. Bull. Fish. Aquat. Sci. 210:1–43; 1981.

Troxler, R. F., A. S. Brown, and S. B. Smith. Bile pigment synthesis in plants. Mechanism of ^{18}O-incorporation into phycocyanobilin in the unicellular rhodophyte, *Cyanidium caldarium*. J. Biol. Chem. 254:3411–3418; 1979.

Chapter 5

Biochemical Features of *Prochloron*

F. Robert Whatley and Randall S. Alberte

Introduction

Prochloron didemni, a photosynthetic symbiont of ascidians, first attracted attention because of its unusual pigment composition (Lewin and Withers, 1975; Newcomb and Pugh, 1975). It contained chlorophylls *a* and *b* but lacked phycobilins (Lewin and Withers, 1975; Thorne et al., 1977), and its ultrastructure revealed a prokaryotic cellular organization (see Swift, this volume). The discovery of *Prochloron* as a prokaryote containing chlorophyll *b* came at a time of accumulating evidence in support of endosymbiotic origins of mitochondria and chloroplasts (e.g., Margulis, 1970; Whatley et al., 1979). Although a number of extant models for the prokaryotic ancestor of the red algal chloroplast, namely, the unicellular cyanobacteria, were already recognized, a potential model ancestor for the green algal chloroplast was missing until the discovery of *Prochloron*. Consequently, many biochemical examinations have been directed toward comparisons between *Prochloron* and cyanobacteria and between *Prochloron* and chlorophyte chloroplasts in attempts to evaluate possible evolutionary relationships.

Most biochemical and physiological investigations of *Prochloron* have been hampered because this alga has not been brought into laboratory culture. It has therefore been necessary to examine material collected in the field, generally in association with colonial didemnid ascidians far from well-equipped laboratories; this means that most biochemical investigations have been carried out elsewhere on frozen or lyophilized material. Even relatively simple experiments to demonstrate the photosynthetic abilities of *Prochloron* cells freshly isolated from their hosts have been complicated by the extreme temperature sensitivity of the alga (Alberte et al., 1986). The discussion here will cover most of the known biochemical features of *Prochloron,* and indicate how biochemical investigations have contributed to an understanding of its phylogenetic relationships. These latter features are summarized in Table 5.1 and Figure 5.1.

General Features of Enzyme Isolation

Until very recently it has proved difficult to extract enzyme proteins from frozen or lyophilized *Prochloron* cells by the usual methods. Fall et al.

Table 5.1 A summary of biochemical features of *Prochloron* indicating closer affinities either to Cyanobacteria (and other prokaryotes) or to green plant chloroplasts or green algae

	Greater affinity to	
Biochemical Feature of *Prochloron*	Cyanobacteria	Green Plant Chloroplast/ Green Algae
Storage carbohydrate	+	+
K_m for RUBISCO	+	–
Presence of chlorophyll *b*	–	+
Absence of phycobiliproteins	–	+
β-carotene content	+	–
Zeaxanthin content	+	–
Absence of allenic carotenoids	+	–
Absence of ε-type carotenoids	+	–
Absence of glycosidic carotenoids	–	+
Carotenoid composition of plasmalemma	+	–
Peptidoglycan wall with muramic acid	+	–
Sterol composition*	–	–
Absence of C_{18} and C_{20} fatty acids	+	–
Abundance of C_{14} and C_{16} fatty acids	+	–
Digalactosyldiacylglycerol content	+/–	–
Sulpholipid content	–	+
Presence of monoglucosyl diacylglycerol	+	–
Absence of polyadenylated RNA	+	–
Absence of ribonucleoproteins	+	+
Absence of small (<5S) RNA species	+	–
DNA base composition (G:C ratio)	+	–

*Sterols derived from animal host.

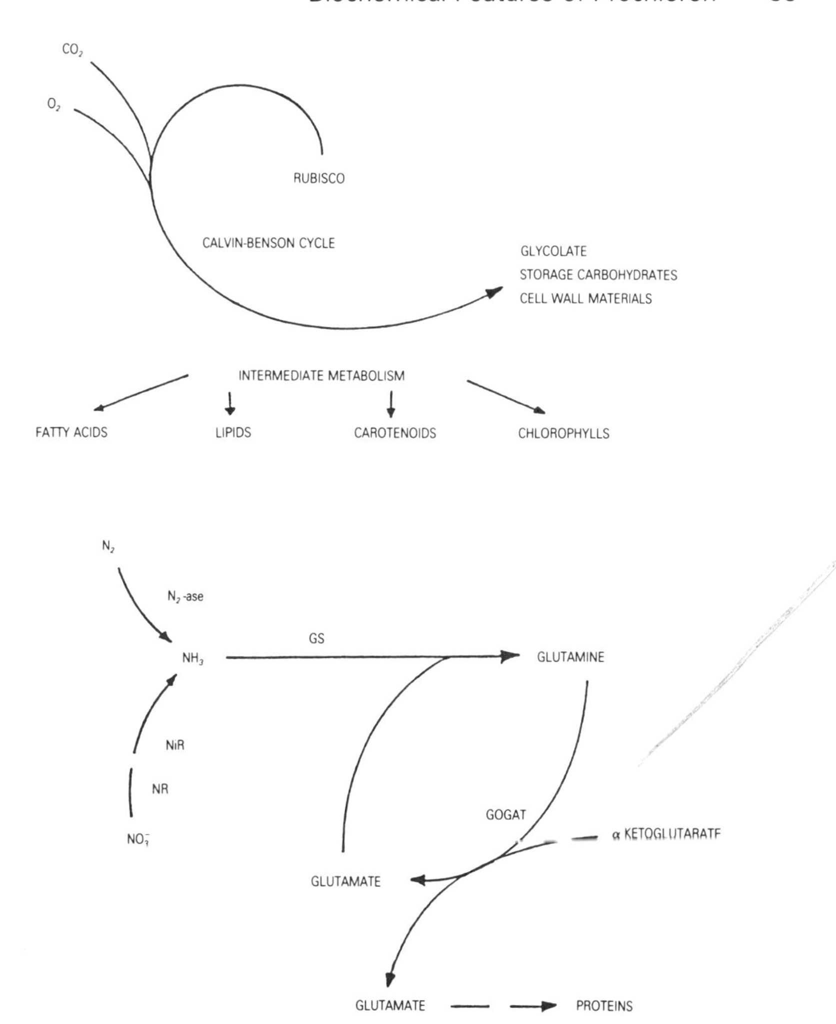

Figure 5.1 A summary of known biochemical pathways and some enzymatic steps in *Prochloron*. Abbreviations: RUBISCO—Ribulose 1,5-bisphosphate carboxylase/oxygenase; N_2ase—nitrogenase; NiR—nitrite reductase; NR—nitrate reductase; GS—glutamine synthetase; GOGAT—glutamine oxoglutarate aminotransferase.

(1983) reported that an extensive "intracellular coagulation" accompanies freezing and suggested this might result from acid released from a storage compartment of the algal cells, or more likely from the didemnid. Subsequent work by Barclay et al. (1987) demonstrated that *Prochloron* cells are rich in phenolics, particularly tannins, which may combine with proteins and other components and precipitate them. Phenolics are known to be very effective inactivators of enzymes and can also lead to coagulation of protein. Additional evidence for the presence of phenolics (or acid) is the ready formation of phaeophytin in suspensions of *Prochloron* cells after brief storage (Thorne et al., 1977).

It was observed that passing cells through a French press did not release active enzyme proteins. Fall et al. (1983) suggested that protein fractions in hypotonic lysates of *Prochloron,* which are easy to prepare at the site of collection, could be precipitated with cold acetone to prepare powders that might prove to contain active enzymes. In fact, glucose-6-phosphate and 6-phosphogluconate dehydrogenase activities were demonstrated in such acetone powders. It is evident from the work of other investigators that some enzymes can be extracted in an active state from frozen material that has been run repeatedly through a French press together with glass beads (e.g., Berhow and McFadden, 1983), and this may be the method of choice in the immediate future. Clearly, one should further evaluate the use of agents (e.g., polyvinyl polypyrrolidone = PVPP) to protect enzymes from inactivation or even precipitation by phenolics. Through the use of PVPP, Alberte et al. (see later) were able to measure the activities of two key enzymes of nitrogen metabolism, glutamine synthetase and nitrate reductase. Activities of a few enzymes from *Prochloron,* including ribulose 1,5-bisphosphate carboxylase-oxygenase (RUBISCO) and phosphoribulose kinase, have been examined in some detail, and some of the enzymes involved in glucan synthesis have been studied to a lesser extent. Most other enzymes have not been studied, and many new areas of metabolism have still to be examined (see Fig. 5.1).

Enzymes of Photosynthetic Carbon Metabolism

Ribulose 1,5–Bisphosphate Carboxylase–Oxygenase

Activity of RUBISCO was first reported in *Prochloron* on the basis of CO_2 fixation experiments with whole cells (Akazawa et al., 1978). Berhow and McFadden (1983) carried out extensive examinations of RUBISCO isolated from frozen *Prochloron* cells. Only three out of seven samples collected from tropical sites yielded active protein extracts, indicating the importance of properly handling the starting material. However, from active extracts they were able to obtain a good purification in a procedure that involved centrifugation in polyethylene glycol and a subsequent sucrose density-gradient centrifugation. The enzyme activity was measured either by incorporating ^{14}C into the acid-stable fraction after incubation with ribulose bisphosphate and $^{14}CO_2$ or by using an enzyme assay that measured phosphoglyceric acid production, coupling its formation with NADH oxidation in the presence of triose phosphate dehydrogenase and phosphoglycerate kinase. Like that of the corresponding enzyme from spinach, the specific activity of *Prochloron* RUBISCO is very low, being approximately 1.5, and the K_m for the substrate, ribulose bisphosphate, is about 9 μM.

Berhow and McFadden (1983) showed that the subunits that could be separated by SDS (sodium dodecyl sulphate) gels were of two sizes, large and small (L = 57.5 kD and S = 18.8 kD), like those from spinach. From the measured native relative molecular weight of 492 k they deduced that *Prochloron* RUBISCO might have a subunit structure of L_8S_8, although their measurements could agree with L_8S_4. Andrews et al. (1984) have also purified RUBISCO from *Prochloron* and compared it with that of a cyanobacterium, *Synechococcus* sp. Their method of purification yielded preparations composed of L and S subunits of 56 kD and 14 kD, respectively. They proposed that RUBISCO from both *Prochloron* and *Synechococcus* sp. has the composition L_8S_8, similar to that in higher plants and green algae. By repeated precipitation of the *Synechococcus* RUBISCO at pH 5.1–5.3, they showed that the S subunits could be removed, resulting in the loss of catalytic activity. With *Prochloron* RUBISCO, this treatment was less effective; when only 22 percent of the S subunits remained, the catalytic activity was reduced to 14 percent. Reconstitution of the stripped (L_8 only) *Synechococcus* RUBISCO by addition of S subunits from the same organism gave fully active homologous enzyme molecules, whereas addition of S subunits from *Prochloron,* although restoring some activity, yielded reconstituted heterologous enzyme molecules with a lower specific activity.

This type of biochemical reconstitution reveals similarities between RUBISCO from *Prochloron* and *Synechococcus.* It would be interesting to know whether S subunits from spinach or green algae could likewise substitute and restore activity to *Prochloron* RUBISCO. Immunological techniques could be used to examine the structural relatedness of *Prochloron* RUBISCO subunits to those of higher plants and other algal and cyanobacterial groups (cf. Plumley et al., 1986). With the current availability of gene sequences and DNA probes for L and S subunits of RUBISCO from various higher plants, algae, and cyanobacteria, it should be possible to obtain nucleotide sequence homologies to *Prochloron* genes and perhaps provide direct evidence of phylogenetic affinities.

Phosphoribulose Kinase

In extracts of *Prochloron,* Berhow and McFadden (1983) detected the presence of phosphoribulose kinase, an enzyme of critical importance to the recycling of carbon intermediates in the Calvin–Benson cycle. Its activity was measured in several different ways, confirming unequivocally the identity of the reaction involved.

Enzymes of Glucan Synthesis and Storage Carbohydrates

Fredrick established (1980) that the starch of *Prochloron* (from *Lissoclinum patella*) contains two α-glucose polymers, (1) a short-chain, unbranched polymer, resembling the long-chain amylose of green algae and land plants but of smaller MW, and (2) a much-branched polymer with α-1,6-linkages, similar to the phytoglycogen of cyanobacteria. (The branched polymers in both *Chlorella* (amylopectin) and the red algae (Floridean starch) have considerably longer α-1,4 chains than phytoglycogen.) *Prochloron* starch thus suggests an intermediate position between cyanobacteria and green algae. In the red algae, storage polysaccharides are synthesized not within the plastid but in the cytoplasm; therefore, the comparison with *Prochloron* is probably spurious.

Using the small amount of frozen *Prochloron* available to him, which necessitated the development of an interesting capillary microelectrophoresis technique, Fredrick (1981) was able to separate and, on the basis of the substrates used, identify two phosphorylase isoenzymes, two glucan synthetase enzymes specific for either ADPG or UDPG, two branching enzymes (which can each make α-1,6 linkages and shorten the chain length of amylose or amylopectin to that characteristic of phytoglycogen), and a glucose-1-phosphatase. Thus, the enzyme armoury of *Prochloron* is demonstrably appropriate for the synthesis of its characteristic starches.

Lipophilic Components

Lipid and Sterol Composition

The lipid composition of lyophilized *Prochloron* cells from *Lissoclinum patella* and *Didemnum molle* (Perry et al., 1978; Johns et al., 1981) included 16 sterols (identified by gas chromatography and mass spectrometry). The three major sterols present (cholest-5-en-3β-ol, 5α-cholestan-3β-ol, and methylcholestan-5,22-dien-3β-ol) were common to both *Prochloron* and one of the ascidian hosts (*Did. molle*). Because of the great similarity in sterol composition between *Prochloron* and *Did. molle,* the authors suggested that the sterols found in the *Prochloron* cell preparations were all derived from the host and were not synthesized by the alga.

Sterols represented about 0.03 percent of the dry weight, whereas fatty acids (5.46 percent) and phytol plus total carotenoids (2.5 percent) were major lipophilic components (Johns et al., 1981). The fatty acid composition of *Prochloron* is thus more like that of cyanobacteria, being dominated by C_{14} and C_{16} acid components (Perry et al., 1978; Johns et al., 1980; Murata and Sato, 1983; Kendrick et al., 1984), whereas C_{18} and C_{20} fatty acids are common to eukaryotic algae (see Kenyon, 1972). The fatty acid fraction included a few polyunsaturated acids. Four major lipid components of *Prochloron* cells from *Dip. similis* were 41 mol% monogalactosyldiacylglycerol (MGG), 4 mol% digalactosyldiacylglycerol (DGG), 47 mol% sulphoquinovosyl diacylglycerol (SL), and 8 mol% phosphatidyl glycerol (PG). Murata and Sato (1983) also detected a small amount of monoglucosyldiacylglycerol in *Prochloron* from *L. patella*. It was pointed out by Kendrick et al. (1984) that the MGG:DGG molar ratio ($\bar{X}$=7.2) is unusually high for photosynthetic cells; most cyanobacteria (Sato et al., 1979) and higher plants have ratios less than 4. In the monoglycosyldiacylglycerol fraction of *Prochloron*, the content of the two major saturated fatty acids, myristic and palmitic, was higher than that in the MGG fraction (Murata and Sato, 1983).

Collectively, the lipid analyses of *Prochloron* show several important features. First, the MGG, DGG, and SL components of *Prochloron* are probably present in both the thylakoid and the plasma membranes, whereas in eukaryotes they are found only in chloroplast membranes. Indeed, no phosphatidylcholine or phosphatidylethanolamine, common features of eukaryotic membranes, has been found in *Prochloron*. Second, the presence in *Prochloron* of monoglucosyldiacylglycerol (Murata and Sato, 1983), an intermediate in the synthesis of MGG in prokaryotic but not eukaryotic algae, suggests that *Prochloron* is more closely related to cyanobacteria than to chloroplasts. Third, the observed variability in the fatty acid composition of *Prochloron* cells collected from different sites and grown under different environmental conditions indicates not only that these features are phenotypically plastic, but also that they should not be used as unequivocal taxonomic markers pending a more thorough study of the effects of environment on lipid composition in *Prochloron*. The fatty acid composition of *Prochloron* may be more related to its marine environment than to its taxonomic position.

Pigments

Chlorophyll *a* and chlorophyll *b* from *Prochloron* were characterized by absorption (Lewin and Withers, 1975) and fluorescence spectrophotometry (Thorne et al., 1977), and the presence of two phaeophytins derived from chlorophylls *a* and *b* was also confirmed. Variations in chlorophyll *a*:*b* ratios observed in *Prochloron* cells isolated from different hosts (Withers et al., 1978; Lewin et al., 1983; Paerl et al., 1984) may be due to differences in light environments and exposure at the collection sites (Bachmann et al., 1985; Alberte et al., 1986, 1987). Typically, the chlorophyll *a*:*b* ratio ranged from 4 to 8 (Lewin et al., 1983; Paerl et al., 1984). The phaeophytins probably resulted from acidification of the cell suspension or from the liberation of phenolics (Thorne et al., 1977). Recently, Burger-Wiersma et al. (1986) reported a new filamentous prokaryote containing chlorophylls *a* and *b* with a ratio of 8:1, rich in β-carotene and zeaxanthin but lacking echinenone, a carotenoid present in *Prochloron* (Withers et al., 1978; Foss et al., 1986).

The phycobilins (allophycocyanin, phycocyanin, and phycoerythrin) are absent (Thorne et al., 1977), and no structures resembling phycobilisomes have been seen in electron micrographs (e.g., Whatley, 1977; Cox, 1986; see also Chap. 7). The associated carotenoids include β-carotene (54 percent) and zeaxanthin (38 percent) as major components (Paerl et al., 1984), with small quantities of echinenone, cryptoxanthin, and isocryptoxanthin, but no glycosidic carotenoids (Withers et al., 1978; Foss et al., 1987). Withers et al. (1978) concluded that this carotenoid composition more closely resembled that of cyanobacteria than that of higher-plant or algal chloroplasts, though the lack of glycosidic carotenoids is a potentially important difference. In addition, the absence of allenic and ε-type carotenoids and the low content of carotenoid epoxides are not compatible with the known pathways of carotenoid biosynthesis in chlorophytes and other algal groups containing chlorophyll *b* (cf., Foss et al., 1987).

Properties of Membrane Fractions

Omata et al. (1985) isolated membrane fractions from living *Prochloron* cells broken in a French press and identified on sucrose density gradients, two membrane fractions that they believe came from plasmalemma or cytoplasmic membranes. The cytoplasmic membrane fraction had a buoyant density of 1.08 g cm^{-3}, close to those from two cyanobacteria and the inner envelopes of pea chloroplasts (buoyant density = 1.13 g cm^{-3}). It contained only a little chlorophyll but had as much zeaxanthin and β-carotene and a few other carotenoids as was reported for whole cells by Withers et al. (1978). In their high carotenoid content the cytoplasmic membrane fractions resembled those of cyanobacteria but differed from the inner membranes of pea plastids, which have a lower total carotenoid content. However, the thylakoid membrane fraction, with a buoyant density of 1.20 g cm^{-3}, was slightly denser than thylakoid fractions obtained from cyanobacteria (buoyant density = 1.18 g cm^{-3}) and spinach chloroplasts (1.17 g cm^{-3}). It contained chlorophylls *a* and *b* together with certain carotenoids, including β-carotene and zeaxanthin, present in much smaller amounts than in the cytoplasmic membranes. In their pigment composition these thylakoid membranes more closely resemble spinach thylakoids than photosynthetic membranes of cyanobacteria.

Examination of the effects of ionic strength and salinity on the photosynthetic activity of *Prochloron* revealed that the plasmalemma of these cells has unusual permeability features (Critchley and Andrews, 1984). *Prochloron* cells isolated from *L. patella* into buffered seawater showed high rates of photosynthesis (see also Alberte et al., 1986); however, when such cells were exposed to isoosmotic solutions of NaCl or distilled water, CO_2-dependent O_2 evolution was severely or completely inhibited. In addition, hypotonic or hypertonic solutions greatly affected the permeability of the plasmalemma to ferricyanide. Collectively, these findings, although not unique to *Prochloron,* suggest the importance of inorganic components of seawater in maintaining membrane properties essential to autotrophic and perhaps symbiotic growth.

Further details on photosynthetic membrane features are treated elsewhere in this volume (see Alberte).

Miscellaneous Compounds and Metabolic Investigations

Proteins and Amino Acids

Omata et al. (1985) extracted unfrozen cells of *Prochloron* from *L. patella* with hypotonic sucrose+Tris buffer and obtained a small amount of a heme-containing compound identified as a water-soluble, c-type cytochrome. It was not further analyzed.

There is a report that *Prochloron* from a number of didemnids (*Dip. virens, Did. molle, Trididemnum* sp., and *L. patella*) contained free 3-(*N*-methylamino)glutaric acid in amounts ranging from 2 to 500 percent of the level found in the glutamate pool of the *Prochloron* (Summons, 1981). Since the algal cells in all cases, except for *Dip. virens* samples, had been freed from the host and its mucilaginous matrix by using 50 percent ethanol, it is likely that this amino acid was, in fact, a constituent not of *Prochloron,* but of the host. In the one case where the *Prochloron* cells had been first separated from the host by release into an aqueous system, the amount of this novel amino acid was extremely low (Summons, 1981).

Cell Wall Components

Moriarty (1979) examined the cell walls of *Prochloron* from *L. patella* and found 1.8 μg muramic acid/mg dry weight of cells. The identification of muramic acid was made by electrophoresis on cellulose and by gas chromatography, and it was quantified by an enzyme-linked assay. Cyanobacteria are known to possess a peptidoglycan (murein) wall made up of alternating units of N-acetylmuramic acid and N-acetylglucosamine, cross-linked by short peptide chains. On the basis of the muramic acid content, equivalent to 1.3 $\mu g \ \mu m^{-2}$ of cell surface, *Prochloron* has a peptidoglycan wall of about the same thickness as that known for cyanobacteria.

Detailed analyses of the peptidoglycan constituents of *Prochloron* isolated from several hosts were made by Stackebrandt and Kandler (1982), who found muramic acid, glutamate, alanine, *meso*diaminopimelic acid, and glucosamine (originating from *N*-acetylglucosamine) to be the major amino acids and amino sugars present. They confirmed the basic peptidoglycan structure of the cell wall, which is composed of carbohydrate chains linked by short peptide chains, and demonstrated that the wall type belongs to the common "Ala-group" of peptidoglycans characteristic of photosynthetic prokaryotes and most Gram-negative bacteria.

Nucleic Acids

Investigations of the nucleic acid composition of *Prochloron* cells from *Did. molle* have shown that the small-molecular-weight ribonucleoproteins and low-*MW* RNAs (<5S) typical of eukaryotic systems could not be detected by electrophoresis (Bachmann et al., 1985). These proteins, usually associated with RNA and forming ribonucleoprotein complexes, could not be found in *Prochloron* even by using sensitive immunological techniques. In addition, poly-A^+ stretches, a common feature of RNAs derived from nuclear genes in eukaryotes and occasionally found in organelle transcripts (Crouse et al., 1984), were not found in the total RNA from *Prochloron* cells when selected on poly(U)-Sepharose.

Herdman (1981) examined the DNA base composition of *Prochloron* and found that it was G-C rich, a feature also typical of cyanobacteria and several green algal chloroplast DNAs (cf. Crouse et al., 1984). However, the G:C ratio of *Prochloron* (40.8 percent) is lower than that found in higher-plant chloroplasts.

Nitrogen Metabolism

Until recently there were no analyses of enzymes associated with nitrogen metabolism in *Prochloron*. Earlier physiological and chemical investigations, however, provided evidence for the existence of certain pathways or activities. Paerl (1984) found high acetylene reduction activity of *Prochloron* in *L. patella* (though not in the algal cells after removal from the host), indicating the presence of an active nitrogenase and N_2-fixing capability. Using ^{15}N–NMR, Parry (1985) was able to demonstrate the synthesis of glutamine from ammonium ions in whole cells of *Prochloron*, implicating a functional GS–GOGAT (glutamine-synthetase–glutamine-oxoglutarate-amino-transferase) system. Surprisingly, this activity appeared to be light-dependent; in chloroplasts the GS–GOGAT system, responsible for NH_3 assimilation, is dependent on light-generated ATP, but in a prokaryotic cell in which the level of O_2 uptake is known to be high (Alberte et al., 1986) this ATP requirement should be met by respiration.

Alberte, Zimmerman, Cheng, and Lewin (in preparation) recently studied the transferase activity of glutamine synthetase (GS) in *Prochloron*. Cell-free preparations of the algal cells from *Lissoclinum patella* exhibited high constitutive activities of this enzyme, not enhanced by feeding ammonium (20 μM) to the whole colonies or to isolated algal cells, and apparently not inhibited by L-methionine–DL-sulfoximine (a known inhibitor of GS). GS activity could be demonstrated only in cells isolated and lysed in buffered seawater (pH above 7.2) containing 1 percent PVPP.

These investigators also demonstrated the activity of nitrate reductase (NR) in *Prochloron* from several host didemnids (*L. patella*, *T. cyclops*, *Dip. virens*, and *L. voeltzkowi*) by using the *in vivo* enzyme assay of Davison and Stewart (1984). Whereas *Prochloron* cells isolated from freshly collected *L. patella* colonies showed little or no NR activity, prior incubation of colonies or isolated *Prochloron* cells with 20 μM nitrate considerably increased the activity over a period of 2–6 h. However, NR activity could not be detected in *Did. molle* colonies, even after they had been similarly incubated with nitrate for as long as 12 h. When ^{15}N ammonium or nitrate was supplied to whole colonies, significant amounts of labeled nitrogen were taken up by the algal symbionts (our unpublished results), indicating that *Prochloron* is normally N-limited, and not (as suggested by Pardy, 1984) adequately supplied by ammonia from the host.

Conclusions

The majority of the biochemical investigations carried out on *Prochloron* (see Fig. 5.1) more closely link this alga with cyanophytes and other prokaryotes, whereas only a few provide ties with green algal or green plant chloroplasts (see Table 5.1). Biochemical investigations alone cannot establish phylogenetic affinities, but they can be expected to reveal biosynthetic and metabolic features of importance to an understanding of symbiosis in general and may perhaps aid in efforts to obtain *Prochloron* in laboratory culture.

References (Additional to Bibliography)

Burger-Wiersma T., M. Veenhuis, H. J. Korthals, C. C. M. Van de Wiel, and L. R. Mur. A new prokaryote containing chlorophylls *a* and *b*. Nature 320:262–264; 1986.

Crouse, E. J., H. J. Bohnert, and J. M. Schmitt. Chloroplast RNA synthesis *in* chloroplast biogenesis (R. J. Ellis, ed.). Soc. Exptl. Biol. Seminar Series 21:83–136; 1984.

Davison, I. R., and W. D. P. Stewart. Studies on nitrate reductase activity in *Laminaria digitata*. Mar. Biol. 77:107–112; 1984.

Kenyon, C. N. Fatty acid composition of unicellular strains of blue-green algae. J. Bacteriol. 109:827–834; 1972.

Margulis, L. Origin of Eukaryotic Cells. New Haven: Yale University Press; 1970.

Plumley, F. G., D. L. Kirchman, R. E. Hodson, and G. F. Schmidt. Ribulose bisphosphate carboxylase from three chlorophyll *c*-containing algae. Plant Physiol. 80:685–691; 1986.

Sato, N., N. Murata, Y. Miura, and N. Ueta. Effect of growth temperature on lipid and fatty acid compositions in the blue-green algae, *Anabaena variabilis* and *Anacystis nidulans*. Biochim. Biophys. Acta 572:19–28; 1979.

Whatley, J. M., P. John, and F. R. Whatley. From extracellular to intracellular: the establishment of mitochondria and chloroplasts. Proc. Roy. Soc., London B204:165–187; 1979.

Chapter 6

Phylogenetic Considerations of *Prochloron*

Erko Stackebrandt

Introduction

Since the first report on *Prochloron didemni* (originally described as *Synechocystis didemni*), a new kind of microscopic alga associated with didemnid ascidians (Lewin, 1975), there have been extensive discussions about its phylogenetic position, its rank, and its role as a possible ancestor of chloroplasts of green algae and higher plants. This prokaryotic phototrophic microorganism, with its unique pigment composition, was discovered in a period when, for the first time in microbiology, it was becoming possible to establish certain phylogenetic relationships and consequently to propose a sound definition of higher ranks of prokaryotes. There were, basically, two different, opposing views. The traditional one bases taxonomic conclusions on comparisons of phenotypic characters (e.g., morphological, biochemical, and physiological properties). On the other hand, the molecular–genetic one determines natural relationships by emphasizing comparative analyses of genomes or parts thereof. Proponents of the first school can in most cases be recognized by their defining *Prochloron* as an alga, prokaryotic alga, or blue-green alga, whereas members of the second school refer to *Prochloron* as a bacterium, a prokaryote, or a cyanobacterium.

I assume that most taxonomists would agree that the taxonomic ranks with their Greek or Latin nomenclature, adopted about 100 years ago by microbiologists from eukaryotic systematics, were established to reflect natural relationships. This means that, prior to their assignment to ranks like classes, divisions, orders, and families, the phylogenetic positions of the respective organisms have to be elucidated. If this is not possible, then the designation and use of suprageneric ranks are worthless because they can be considered to contribute to confusion instead of shedding light on the evolution of organisms. At present, the phylogenetic tree of prokaryotes (including cyanobacteria) is far from established, but its outlines are just emerging. It is not rash to prophesy that a hierarchic system for prokaryotes will soon be developed, since methods to elucidate the phylogenetic position of any organism are either in use already or will be developed in the near future. However, the situation today is somewhat confused and illogical, because we currently recognize certain taxa for which the phylogenetic position and coherency have been demonstrated, whereas other taxa, whose ranks may or may not reflect their actual phylogenetic position, are defined purely phenotypically, and thus may comprise a heterogeneous collection of phylogenetically diverse organisms.

Phylogenetic Position

The phylogenetic position of *Prochloron didemni* is not yet settled. There is no doubt that all samples isolated from different didemnid ascidian hosts are closely related, as indicated by DNA : DNA reassociation studies on seven strains (Stam et al., 1985) and rRNA cataloguing data from four strains (Stackebrandt et al., 1982). Whether or not to describe these isolates as separate strains of *P. didemni,* or as representing individual species, is largely a matter of taste, because there are so few distinguishing characters. There is also little doubt that *Prochloron* shares a closer relationship to cyanobacteria and to chloroplasts of red algae, green algae, *Euglena,* and higher plants than to any other prokaryote. Cyanobacteria, *Prochloron,* and all chloroplasts investigated so far form one of the major branches of the eubacterial tree. On this branch the cyanobacteria, along with the chloroplast of *Porphyridium,* form a coherent cluster, whereas all other chloroplasts branch off elsewhere. We now have analyses of 5S rRNA obtained by complete sequencing (Mackay et al., 1982), 16S rRNA obtained by conventional average linkage clustering of similarity coefficients (Seewaldt and Stackebrandt, 1982), and additional information obtained by reverse transcriptase sequencing (Giovannoni *et al.**).

*Giovannoni et al., Poster at the XIV International Botanical Congress, 1987, Berlin (West), F.R.G.

Collectively, these data indicate that *Prochloron* probably falls within the realm of the cyanobacteria. This relationship is in agreement with data on the DNA-base composition and genome size of the one *Prochloron* strain investigated so far, which fall within the ranges reported for cyanobacteria (Herdman, 1981). However, dendrograms of relationships are valid indicators of phylogenetic distances only if the organisms investigated have evolved isochronally, which is hard to prove by 16S rRNA cataloguing alone. Even van Valen (1982), in his redesigned tree (which erroneously compares data based on dissimilar methods of analysis; Bandelt, pers. communication), places *Prochloron* in the cyanobacteria–chloroplast cluster. Even if the *Prochloron* genome had evolved from those of cyanobacteria (which will probably be shown when we have more data on full sequences of conserved genes), the phylogenetic position of *Prochloron* will not be very different from that proposed today. Lewin (1983, 1984, 1986), in discussing various possibilities about the origin of *Prochloron*, suggested that a unicellular, marine cyanophyte could have been the ancestor of the didemnid ascidian symbionts. I agree that this seems plausible, but an understanding of the exact order of events leading to the evolution of *Prochloron* has to await new insights, from other data, at different levels of genome and cell organization.

There is a similar degree of uncertainty about the epoch in which *Prochloron* might have evolved. The fossil record gives us no indications, and the finding that *Prochloron* can be "fossilized" in the laboratory is of little significance, since this can also be done for cyanobacteria (Francis et al., 1978). Lewin (1986) suggested that the time for the origin of *Prochloron* was some 100 million years ago. The phylogenetic data are not able to give a precise answer, though we can conclude that the branching of *Prochloron* from the cyanobacteria happened less than 1.8 billion years ago.

Phylogenetic Rank

The second problem is connected with the assignment of ranks. There is no question that the genus *Prochloron* is a valid one. The main problem is where to place it. So far the descriptions of higher ranks among algae and related microbes have been based on the assumption that their unique pigment compositions are of primary phylogenetic importance. This puts *Prochloron* in an order Prochlorales, equivalent to that of Cyanobacteriales, in subclass I, Oxyphotobacteria, of the class Photobacteria (Gibbons and Murray, 1978). (Although these authors state that all members of this subclass are prokaryotes, neither *Cyanobacterium* nor *Prochloron* is included in the Approved Lists of Bacterial Names. *Prochloron* has recently been described validly in accordance with the International Code of Nomenclature of Bacteria [Florenzano et al., 1986], and we hope that a valid description of *Cyanobacterium* under the Bacteriological Code will follow.) If future work supports the earlier indications that *Prochloron* evolved within the phylogenetic confines of cyanobacteria (tentatively designated Cyanobacteriaceae), then it has to be considered as a separate genus of this family. The presence of chlorophyll *a* and *b*, and the lack of phycobiliproteins, would consequently necessitate an emendation of the family and order diagnoses (Antia, 1977). However, no rank higher than a genus could be assigned to such microorganisms as *Prochloron*. If, on the other hand, *Prochloron* (and its allies, if any) and cyanobacteria (as defined today) share a common ancestry and evolved independently from each other, then both groups are taxonomically equivalent, as expressed today in their assignment to different orders of the same subclass. The most disturbing discrepancy seen today is the allocation of the orders Cyanobacteriales and Prochlorales to the class Photobacteria (Gibbons and Murray, 1978; Florenzano et al., 1986). Here again the establishment of higher ranks does not reflect the phylogenetic relationships of the organisms involved. The class Photobacteria, as originally defined, should include only phototrophic organisms, which is definitely not the case. Most "orders" of phototrophic organisms also comprise nonphototrophic relatives (or *vice versa*, which makes most of the ranks described by Gibbons and Murray [1978] meaningless).

Possible Relationship to Chloroplasts

The third problem is connected with the role *Prochloron*-like microbes might have played as the ancestors of chloroplasts of green algae and higher plants. Nothing is known about a possible independent evolution of chlorophyll *b* and the formation of a functional chlorophyll (*a* + *b*)–protein complex. The finding that this complex differs somewhat in size and nature from that present in green algae and higher plants (Schuster et al., 1984; Hiller and Larkum, 1985) may be interpreted in terms either of differences in the rates of evolution of the respective genes or of their distinct origins. One may speculate about the loss of genes for phycobiliprotein synthesis as a consequence of the evolution of a more effective chlorophyll (*a* + *b*)–protein complex (Giddings et al., 1980). The possibility of a "lateral" gene transfer to account for the occurrence of chlorophyll *b* in *Prochloron* has been discussed by Lewin (1986).

Referring to the phylogenetic data again (MacKay et al., 1982; Seewaldt and Stackebrandt, 1982; Herdman, 1981; Giovannoni and co-workers, unpublished), a specific relationship to the chlorophytes and the proposed allocation of *Prochloron* to a new class of Chlorophyta (Tseng and Zhou, 1983) can be definitely excluded. It must, however, be kept in mind that the genomes of many chloroplasts seem to have been evolving faster than those of the cyanobacteria, which makes a direct comparison difficult. As long as the monophyletic origin of chlorophyll *b* biosynthesis has not been demonstrated, one should argue carefully, perhaps following the suggestions of Chadefaud (1978, Figure II), who postulated independent mutations for the origin of chlorophyll *b* in *Prochloron*, Euglenophyceae, and Chlorophyceae and proposed that *Prochloron* had originated from a cyanobacterial ancestor.

References (Additional to Bibliography)

Gibbons, N. E., and R. G. E. Murray. Proposals concerning the higher taxa of bacteria. Int. Journal of Systematic Bacteriology 28:1–6; 1978.

Chapter 7

The Cytology of *Prochloron*

Hewson Swift

Introduction

Cytological observations made on *Prochloron* cells from a variety of didemnid hosts present a consistent picture. Characteristic differences occur between cells from different sources (Figures 10, 12–14), but obvious similarities indicate that these organisms form a coherent and probably closely related group. Prochlorophytes are clearly prokaryotes. Nuclei and mitochondria are absent. Photosynthetic membranes are free in the cell and are not packaged into chloroplasts. Various inclusions, DNA filaments, protein crystalloids, pigment and storage granules are usually not membrane-limited. The only other true membrane systems present are the cell-limiting plasmalemma and membranes surrounding vacuole-like regions of low optical density.

In some characteristics, *Prochloron* cells resemble blue-green algae. The cell wall is multilayered, usually with a fibrous outer sheath. Cell division is by cross-wall partitioning, apparently without a specific mechanism for genome segregation. Polyhedral bodies are frequently present, and have been shown to contain ribulose 1,5-bisphosphate carboxylase (Swift

The author wishes to thank Ralph Lewin and Lanna Cheng for much advice and guidance concerning the appearance and distribution of *Prochloron* and its didemnid hosts among the coral shoals of Palau, and Randall Alberte, who first introduced him to this interesting group of organisms. The technical help of Dr. Chris Chow and Mrs. Sagami Paul is gratefully acknowledged. Original work reported here was aided by Grant CA-14599 from the National Institutes of Health, Grant N00014-88-K-0258 from the Office of Naval Research, and the Kiwanis International.

and Leser, 1989). These are similar to the carboxysomes of cyanobacteria and chemoautotrophic bacteria (Shively et al., 1973; Codd and Marsden, 1984). Thylakoid structure, however, clearly distinguishes *Prochloron* from blue-green algae. (Thylakoids are flattened sacs of photosynthetic membranes.) In cyanobacteria the two *inner* surfaces (figures 15 and 16) are often tightly appressed, and only occasionally is the intrathylakoidal space evident (Edwards et al., 1968); phycobilisomes cover the outer surface of the thylakoids, which are characteristically widely spaced, and only rarely are the membranes of two separate thylakoids in contact (Gantt and Conti, 1969), although in developing or degenerating cells membrane stacks may occasionally be evident (Echlin, 1964). By contrast, in *Prochloron* the intrathylakoidal space is usually apparent, and many thylakoid membranes are fused by their *outer* faces with other thylakoids. Thylakoids thus occur in pairs or stacks. This fusion pattern is, of course, also characteristic of thylakoids in eukaryote chloroplasts, although the formation of grana, as in higher plants, is not evident in *Prochloron*. A similar fusion of membranes between the outer surfaces of two flattened vesicles is also characteristic of certain chemoautotrophic bacteria (e.g., in *Nitrosocystis oceanus* and *Rhodobacter viridis* [Watson and Remsen, 1970; Miller and Jacob, 1985]).

In most bacteria and blue-green algae the DNA-containing filaments are centrally located in the cell in one or only a few nucleoids. In *Prochloron*, as in most chloroplasts, the distribution of DNA is much more diffuse, with multiple centers of concentration usually around the periphery (Coleman and Lewin, 1983).

In almost all instances *Prochloron* cells occur in intimate but extracellular association with their didemnid hosts, either attached to the outer surface or lying within the atrial cavity where they are bathed in the water currents circulated by the cilia of the branchial sac (Newcomb and Pugh, 1975; Cox and Dwarte, 1981; Cox, 1986). In some tissue preparations algal cells can be seen in the branchial stigmata, in pockets of the atrium (Pardy et al., 1983), and occasionally in the branchial cavity (Cox, 1983). (See Figures 1, and 2.) A few *Prochloron* cells may be engulfed by host phagocytes (Cox, 1983) and cell fragments may infrequently be seen within the digestive system. Isolated cells occasionally appear surrounded by host extracellular matrix, but careful analysis shows often they are within a cavity outlined by the very thin squamous epithelial layer of the host, and thus lie in a tubular outpocketing of the atrium. *Prochloron* cells from *Didemnum carneolentum* (=*Did. candidum*?) are concentrated in feeding grooves on the outer surface of the colony (Lewin, 1975, 1977, 1981; Whatley, 1977), and similar surface association with ascidians of three other families (the Polycitoridae, Polyclinidae, and Styelidae) have also been reported (Cox, 1986). Rare associations with other hosts, namely on an encrusting coralline alga (Lewin, 1981), on outer surfaces of the holothurian *Synaptula lampertii* (Cheng and Lewin, 1984), and in the sponge *Aplysilla* (Parry, 1986), need further observations before they can be considered to be more than adventitious.

Sexual reproduction in didemnids, as in other ascidians, involves fertilization and development of eggs in diverticula of the atrium, the mature tadpole-shaped larva emerging through the atrial siphon. In *Lissoclinum patella, Didemnum molle,* and certain other species (Eldredge, 1965), the tadpoles acquire a sheath of symbiont cells, adhering to epidermal folds in their lateral and ventral surfaces (Figures 3, 4 and 5, see also Kott, 1981). In the formation of a new colony these larvae thus take their symbionts with them. Where the association with the host is external and facultative, however, the larvae are not specially modified for carrying *Prochloron*. Apparently in these host species, new colonies are infected by symbiont cells present with other microplankton in the sea water.

The Cell Wall

Cell walls in most preparations of *Prochloron* cells closely resemble the walls of cyanobacteria and Gram-negative bacteria (cf. Figures 16 and 29). *Prochloron* cells from *Didemnum carneolentum* (Schulz-Baldes and Lewin, 1976) and *Diplosoma virens* (Whatley, 1977) show similar patterns of wall structure. The total cell wall material, including plasmalemma but not outer sheath, is 35–50 nm thick. There are two electron-dense layers outside the plasmalemma, separated by layers of lower density. The outer dense layer is wavy. The cell wall of algal cells from *Didemnum molle* (Cox and Dwarte, 1981) also has two dense outer layers, but these are somewhat thinner and not wavy, and lack sheath filaments. Some of the reported differences may be associated with variations in specimen preparation. In our material of cells from both *L. patella* and *Did. molle* (Figures 28 and 29), sheath components, fine filaments and club-shaped processes are evident.

In our light-microscope studies, the cell walls of *Prochloron* from *L. patella* apparently contain polysaccharides, as evidenced by a strongly positive periodic-acid-Schiff (PAS) reaction. They are also basophilic and weakly metachromatic (RNase insensitive) and thus evidently contain acid groups, probably sulfate esters. Basic dye staining reveals a sulcus of thinner wall material surrounding many cells (Figure 8). In dividing cells this sulcus appears to lie perpendicular to the division plane (Figure 32), and thus is not a part of the partitioning septum that forms before cell division. When fresh cells from *L. patella* are lysed by osmotic shock, clam-shell-like cell wall ghosts are formed, presumably because the cell wall ruptures along the sulcus to release the cell's contents (Figures 33, 34 and Color Plate 2.30). *Prochloron* cells from *Did. molle* lack an obvious sulcus, and in scanning electron micrographs appear to have a quite different texture (Figures 30 and 31). In sectioned material other differences are evident, as described in the captions to Figures 28 and 29. Cell division, as in many other prokaryotes, is preceded by a transverse septum that forms as the cell shape changes from spherical to oblong (Figures 17 to 20). The septum begins as an inward extension of the cell membrane, in association with the inner dense wall layer and the low-density intermediate layer. The outer dense layer may bulge outward at this point. As the septum grows inward it is often lined by a thin layer of cytoplasm between the septum and central vacuole (Figure 20).

Thylakoids

As pointed out in all structural studies on *Prochloron*, photosynthetic lamellae form a dominant feature of the cell, but their arrangements differ in cells from different hosts, as shown in Figures 10, 12–14 (Newcomb and Pugh, 1975; Cox, 1986). The thylakoids in algae from *L. voeltzkowi, L. patella, Diplosoma virens,* and *Trididemnum cyclops* are usually arranged in concentric layers around a large central vacuole limited by a membrane apparently contributed by the thylakoids. The central vacuole often appears "empty" in light (Figure 1) and electron micrographs, the contents presumably leached during specimen preparation. In other cells, it may contain flocculent material (Figure 13) or membrane fragments (Figure 21). When fresh algal cells from *L. patella* are lysed with hypotonic medium, the central vacuole may be extruded intact to appear under the phase microscope as a uniformly gray sphere. In cells from *L. patella* the vacuole comprises a variable fraction of the cell volume and may even be absent altogether. As pointed out by Cox and Dwarte (1981), ribosomes are excluded from the thylakoid layers, which alternate with stromal regions of the cell rich in ribosomes (Figure 11). Thus, when thin sections are stained for RNA (as in Figure 8), arcs of stained and unstained regions are evident. Although the number of thylakoids in a stack varies from place to place within a single cell, there are differences in stack size associated with *Prochloron* from different hosts. In *Prochloron* cells from *Dip. virens* many lamellae contain only two thylakoids as in Figure 12 (Whatley, 1977), whereas algal cells from *L. patella* may have domains of paired thylakoids interspersed with much thicker stacks. More than 30 appressed thylakoids are evident in Figures 13 and 21, but in many regions thylakoids are only in groups of two (Figure 23). The claim by Thinh (1978) that some *Prochloron* cells from *Dip. virens* "possess single non-appressed thylakoids" seems to be a misinterpretation, based on material in which thylakoidal structure was swollen and indistinct.

Although most algal cells from *L. voeltzkowi*, *L. patella*, and *Dip. virens* have a large central vacuole, algae from *Did. molle* usually lack such a central vacuole, and thylakoids and ribosome-containing regions have a much less orderly arrangement throughout the cells (Figures 7 and 13) (Cox and Dwarte, 1981; Cox, 1986).

Single thylakoid membranes are about 8 nm thick, and the tightly fused paired membranes are about 15 nm. The fusion produces a characteristic five-component membrane complex of three dense lines separated by intervening low-density regions. The middle element, representing the fused outer surfaces of two thylakoids, is particularly electron–dense (Figures 22 and 23), as it is also in fused thylakoid membranes of eukaryote chloroplasts. This density is not due to staining with osmium alone, since it is evident also in cells fixed only in glutaraldehyde.

Freeze-cleaved preparations have been made in studies of *Prochloron* thylakoids from two different hosts. Giddings et al. (1980) studied the distribution of protein particles in algal cells from *Dip. virens,* reporting in the (exoplasmic) E face (abbreviated as EF) a particle density in fused membranes more than three times as great as in unfused membranes. Fracturing divides the single thylakoid membrane into two components. The EF represents the newly exposed face of the inner component, which lines the intrathylakoid cavity. Measurements of particle diameters show size category maxima at 7.5, 10.5, 13.0, and 16.0 nm. These values closely match the sizes found in fused thylakoid membranes of *Chlamydomonas* and pea chloroplasts, thought to represent the reaction centers of photosystem II with 0, 1, 2, or 4 units of the chloroplast ($a + b$) light-harvesting complex. Unstacked membranes show only a single class of EF particles averaging 8 nm. Micrographs of the freeze-cleaved replicas of *Prochloron* cells from *Dip. virens* (Giddings et al., 1980) are shown in Figures 24–27. Where the fracture plane crosses two unstacked membranes the membrane separation is evident as a densely shadowed line (shown between arrows in Figure 27). Where thylakoid membranes are in contact within the stacked regions, the boundaries between two membranes are less marked. Thus, the regions where membranes are unstacked or stacked are distinguishable on the replicas.

Measurements by Cox and Dwarte (1981) of freeze-cleaved thylakoid membranes of *Prochloron* from *Did. molle* are less extensive and less clear, partly because unstacked thylakoids are scarcer in this material. Stacked thylakoids bear EF particles of 15.6 nm, but also contain regions where the particles are smaller (13.2 nm) and more sparsely distributed. Particles from the (protoplasmic) P face of fused membranes are, on the average, slightly larger, 12.7 nm, with maxima at 10 and 15 nm, as compared with 12.0 nm for particles on the PF of unfused membranes.

In the profusion of membranes that often form irregular or sinuous patterns in sections of the *Prochloron* cell, it may be difficult to determine whether a particular space or inclusion lies in or adjacent to a thylakoid. Cox and Dwarte (1981) concluded that the carboxysomes are bounded by specialized thylakoid membranes with particle distributions different from those in the typical thylakoid stacks. In some cells (e.g., in *Prochloron* from *Dip. virens* [Withers et al., 1978]) typical carboxysomes apparently lie in the ribosome-rich areas, outside the thylakoids. Even in cells with a well-defined central vacuole, the vacuolar region seldom seems to be outlined by a single, clearly continuous, distinct membrane, like that around the central vacuole (tonoplast) of many cells of higher plants. Whatley (1977) pointed out that the central domain frequently appears to be bounded only by thylakoids, as evident in Figures 10 and 12. The numerous smaller vacuoles present in cells without an obvious central region may also be limited by thylakoid membranes. It would be of interest to apply the photosynthetically linked diaminobenzidine (DAB) reaction (Lauritis et al., 1975) to these cells to determine whether all limiting membranes in the cell are potentially active in photosynthesis or whether some membranes might be specialized for other, more structural roles.

Much further study of thylakoid structure in prochlorophytes is required before the nature of their complex architecture in relation to PSI and PSII becomes clear. The particle measurements of Giddings et al. (1980) led them to the tentative conclusion that the thylakoid stacking in *Dip. virens* closely parallels the stacking of grana in higher-plant chloroplasts, with segregation of PSII components in stacked regions and PSI components in unstacked stromal membranes. Thus, the stacking of prochlorophyte membranes is seen as a true functional organization, quite different from the unstacked, widely spaced phycobilisome-bearing thylakoids of cyanobacteria.

Inclusions

At present our knowledge of inclusions in *Prochloron* cells is based largely on their morphology and on one immunocytochemical study. Adequate functional characterization must await the time when cells can be readily cultured in the laboratory to provide sufficient quantities for cell fractionation. Most investigators have described polyhedral bodies in *Prochloron* cells from all hosts studied so far. These have been shown to contain ribulose 1,5-bisphosphate carboxylase–oxygenase (RUBISCO) in cells from *L. patella* and *Did. molle*. We demonstrated this immunochemically by utilizing a polyclonal antibody raised against the large RUBISCO subunit from higher plant chloroplasts (Swift and Leser, 1989). They can thus properly be called carboxysomes (Figure 37). They closely resemble the carboxysomes of cyanobacteria and autotrophic bacteria that have also been shown to contain RUBISCO (Shively et al., 1973; Codd and Marsden, 1984). They are usually 0.2 to 1 μm in diameter, are of medium to high electron density, generally showing no resolvable internal crystalline substructure, but have a crystalloid (polygonal) shape—square, hexagonal, or pentagonal in outline, often with somewhat rounded corners. They frequently occur in clusters of two or more. Examples are evident in micrographs published by Schulz-Baldes and Lewin (1976), Whatley (1977), Withers et al. (1978), Griffiths et al. (1984), and Thinh et al. (1985). In our material (Figures 10, 11, 13, 14, 15, 17, and 37) no limiting membranes are apparent, although a carboxysome may occasionally indent a thylakoid and thus appear partially membrane-bound. In some cyanobacteria, carboxysomes have electron-dense margins, which may be a fixation artifact as discussed by Codd and Marsden (1984), but no true membrane.

Other protein crystals, some of which are large enough to be readily visible by light microscopy, are clearly of several types. In *Prochloron* cells from *Trididemnum cyclops* Griffiths et al. (1984) described square crystals, sometimes more than 3 μm on a side, with an internal lattice spacing of 14–18 nm. Thinh et al. (1985) also described larger crystals, up to 8 μm long with parallel sides but oblique (hexagonal) end faces, with a particle–lattice spacing of 9–10 nm, and other crystals with hexagonally packed subunits showing a spacing of about 29 nm. Somewhat similar crystals were found in *Prochloron* from *Did. molle, Did. sp., L. patella, L. punctatum,* and *Trididemnum miniatum*. Some of the apparent variability of lattice spacing is clearly associated with the angle of sectioning, but, as the authors point out, the range in spacings from 9 to 29 nm is too large to be produced by differences of sectioning angle alone. Studies with a goniometer (tilting) stage and involving optical diffraction of micrographs are needed to help us classify the number of crystal types actually present. Typical examples of crystal structure in algal cells from *L. patella* and *Did. molle* are shown in Figures 10, 28, 35 and 36.

Griffiths et al. (1984) and Thinh et al. (1985) also encountered a variety of paracrystalline arrays in *Prochloron* from several different hosts. These are small aggregations of dense subunits grouped in lattice-like arrangements of rods or evenly spaced granules. Some of these structures are seen adjacent to polyhedral bodies. A variety of other inclusions, including round bodies, have been reported in algal cells from *Didemnum sp.*, containing concentric lamellae somewhat like those in the lipoprotein inclusions of some eukaryote lysosomes. Also, in cells from *Did. molle,* clusters of osmiophilic granules, angular in outline, have been described. Clear vesicles, apparently partially leached in processing, with traces of dense material remaining at their periphery, are probably polyphosphate storage bodies, since they are PAS-positive and metachromatic, and thereby resemble polyphosphate inclusions of yeast and other microorganisms (Figure 10). Similar bodies have been described by Whatley (1977).

In our studies on algal cells from *L. patella,* the frequency of large crystalloids was readily determined in unstained whole cells viewed by polarizing microscopy. There was, on average, about one large crystal per cell, although some cells had several and others none. Slightly less than half of the crystals were naturally colored bright orange red. These remained intact in cells disrupted with sodium dodecyl sulfate and protease K. The pigment was partially soluble in acetone, with absorption maxima at 495 and 540 nm. It is probably a carotenoid. In unstained preparations of algal cells from *Did. molle* such large crystals were absent, but all cells had numerous small birefringent crystals, some of which were probably polyhedral bodies, that gave them a sparkling, jewel-like appearance when viewed with polarized light. [It seems unlikely that these fine crystals were derived from the salts used in tissue preparation, since in *Did. molle* material, embedded in glycolmethacrylate and cut at 1 μm for light microscopy, one could make the *Prochloron* cells stand out clearly from host tissues by crossing the polars, which made the highly birefringent algal cells appear to glow, while host tissues remained dark.]

Many *Prochloron* cells from *Did. molle* contain red-brown amorphous pigment granules. It seems likely that some of the osmiophilic material described in cells from *Did. molle* by Thinh et al. (1985) represents similar granules. It is also possible that some of these granules were confused with carboxysomes by Cox and Dwarte (1981), who described the carboxysomes in cells from *Did. molle* as unusually numerous and electron lucent if osmium tetroxide was not used for fixation. In our cytochemical studies (Swift and Leser, 1989) of this species we observed only a few carboxysomes per cell, with the expected electron density (Figures 14 and 37). The pigment granules, however, were osmiophilic and very numerous. [Compare the density of osmium-fixed granules (Figures 14 and 36) with those fixed only in glutaraldehyde (Figure 37).] Other *Prochloron* cells from the same host are bright green and contain no visible brown pigment. Since cells with various intermediate levels of brownish pigment were also seen, it seems unlikely that these represent two separate species. In fresh preparations from Palau, the green cells are grouped in the corners of the didemnid atrium; they seem somewhat more fragile than the brown cells, lysing more easily in hypotonic media. Since *Did.*

molle has in its outer test a layer of pigment cells, also colored a reddish brown, it seemed possible that the algal cells may somehow acquire pigment from their host. However the melanocytes of *Did. molle* have pigment granules that are only weakly basophilic, whereas in *Prochloron* the pigment is strongly basophilic and metachromatic. No similar pigment is seen in host cells, although extracellular masses of pigment occur in pockets of the atrium, outside *Prochloron* cells, possibly originating from lysed or damaged algae.

The role of this brown pigment in *Prochloron* from *Did. molle* is presently very unclear. It may be present in only one stage of the life of the cell.

Nucleic Acids

Most electron-microscope studies on *Prochloron* have revealed aggregates of small, dense particles 1.5 to 2.0 nm in diameter, described as ribosomes (Figure 11). In thin sections of cells with well-defined central vacuoles and extensive stacks of thylakoids, as shown in *Prochloron* from *L. patella*, ribosomes are apparently excluded from the regions of photosynthetic lamellae, and are limited to bands arranged in concentric arcs around the vacuole. This is clearly demonstrated also in cells embedded in glycolmethacrylate, cut at 1–2 μm, stained for RNA with Azure B, and examined by light microscopy (Figure 8). In algal cells from *Did. molle*, which lack a central vacuole and where thylakoid stacks are smaller and less regular, the RNA staining is more diffuse.

Fine filaments, thought to be fibers of DNA, have been reported by Schulz-Baldes and Lewin (1976) and Thinh (1979). The filaments described by Schulz-Baldes and Lewin were located in an area of low density, surrounded by regions of granular material similar to the ribosome-containing bands described earlier. The filaments reported by Thinh (1979) were present in the central vacuole of the cell. In our electron-microscope studies on *Prochloron* cells from *L. patella*, DNA filaments were evident in material fixed in glutaraldehyde (Figure 11). In material fixed in this way enzymatic extraction of DNA was incomplete, but the filaments could be completely dissolved by hot trichloroacetic acid, under conditions that rendered negative the DAPI staining of DNA under the light microscope (as explained later). Although the DNA filaments were more concentrated in some regions than in others, they formed a network in all the ribosome-containing bands throughout the cell, but were absent from the central vacuole. This diffuse distribution is quite unlike the usual central disposition of DNA strands in unicellular or filamentous blue-green algae. It is quite similar, however, to the distribution of DNA in the chloroplasts of higher plants (see, for example, Kislev et al., and Swift [1965]).

The localization of very small quantities of DNA by light microscopy is possible by utilizing the intense fluorescence of diamidino-phenylindole or DAPI. With this stain, Coleman and Lewin (1983) described the DNA-containing regions of *Prochloron*, from five different hosts, as occurring in 15–50 irregularly shaped aggregates, up to 5 μm long, scattered among the thylakoids at the cell periphery.

In our studies, DAPI staining of DNA was somewhat masked by RNA binding, as well as by the strong autofluorescence of unidentified components in the cell. DAPI fluorescence was best evaluated in ribonuclease-extracted glycol-methacrylate sections, or in whole cells flattened on a glass slide. Algal cells from *L. patella* showed a ramifying network of fluorescent material, with bright centers interconnected by fine anastomosing strands, an arrangement that precisely paralleled the distribution of filaments seen by electron microscopy (Figure 9; Color Plate 2.29). *Prochloron* cells from *Did. molle* seemed to possess a similar distribution of DNA, although the bright centers were larger and fewer, but the DNA staining was difficult to evaluate because of the strong autofluorescence (Swift and Leser, 1989). One needs to develop methods for extracting or quenching the autofluorescence.

Conclusion

Cytological studies to date have given us adequate descriptions of the major characteristics of *Prochloron* from a few didemnid hosts, and glimpses of the structure of a few more. Basic aspects of cell wall, thylakoid, and "cytoplasmic" components have been described, but much more work needs to be done (e.g., on the chemical identification of specific cell inclusions and on their roles in cell metabolism). It now seems clear that there are characteristic morphological differences between *Prochloron* cells from different hosts, worthy of at least species distinction, although much more knowledge is required before adequate taxonomic characters are established.

The chlorophyll content and other aspects of *Prochloron* photosynthesis suggest that distant prochlorophyte ancestors may have played a pivotal role in the evolution of eukaryotic green plants. To the theory of the symbiotic origin of chloroplasts, cytological studies such as these can add evidence for an essentially chloroplast-like arrangement of paired and stacked thylakoids, a size distribution of integral proteins similar to that found in higher plants, and a dispersed chloroplast-like distribution of DNA.

References (Additional to Bibliography)

Codd, G. A., and W. J. N. Marsden. The carboxysomes (polyhedral bodies) of autotrophic prokaryotes. Biol. Rev. 59:389–422; 1984

Echlin, Patrick. The fine structure of the blue-green alga *Anacystis montana f. minor* grown in continuous illumination. Protoplasma 58:439–457; 1964.

Eldredge, L. G. A taxonomic review of Indo-Pacific didemnid ascidians and descriptions of 23 Central Pacific species. Micronesica 2:161–261; 1965.

Gantt, E. and S. F. Conti. Ultrastructure of blue-green algae. J. Bacteriol. 97:1486–1493; 1969.

Kislev, Naomi, Hewson Swift, and Lawrence Bogorad. Nucleic acids of chloroplasts and mitochondria of Swiss chard. J. Cell Biol. 25:327–344; 1965.

Kott, P. Didemnid-algal symbioses: algal transfer to a new host generation. Proc. Fourth International Coral Reef Symposium, Manila. 2:721–723; 1981.

Lauritis, James A., Eugene L. Vigil, Louis Sherman and Hewson Swift. Photosynthetically linked oxidation of diaminobenzidine in blue-green algae. J. Ultrastr. Res. 53:331–344; 1975.

Miller, K. R., and J. S. Jacob. The *Rhodopseudomonas viridis* photosynthetic membrane: arrangement in situ. Arch. Microbiol. 142:333–339; 1985.

Shively, J. M., Frances Ball, D. H. Brown, and R. E. Saunders. Functional organelles in prokaryotes: polyhedral inclusions (carboxysomes) of *Thiobacillus neopolitanus*. Science 182:584–586; 1973.

Swift, Hewson. Nucleic acids of mitochondria and chloroplasts. Amer. Naturalist. 99:210–227; 1965.

Swift, H., and G. Leser. Cytochemical studies on prochlorophytes: localization of DNA and ribulose 1,5-bisphosphate carboxylase-oxygenase. Phycologia (in press); 1989.

Watson, Stanley W., and Charles E. Remsen. Cell envelope of *Nitrosocystis oceanus*. J. Ultrastr. Res. 33:148–160; 1970.

Captions For Figures 1–37

Figures 1–5. Light micrographs of *Prochloron* cells in association with their didemnid ascidian hosts. Tissues have been fixed with glutaraldehyde and osmium tetroxide, embedded in epon, sectioned at 1 μm and stained with toluidine blue at pH 8.0.

Figure 1. Portions of two zooids of *Lissoclinum patella* surrounded by *Prochloron* cells outside the pharyngeal basket and within the communal atrial sinus. About 200 X.

Figure 2. A zooid of *Lissoclinum voeltzkowi,* as above, with *Prochloron* cells crowding the atrial sinus. About 200 X.

Figure 3. Portion of a tadpole larva of *L. patella* at the left, showing *Prochloron* cells attached to folds of the ventrolateral surface. About 300 X.

Figure 4. Enlarged portion of Figure 3. 750 X.

Figure 5. Live tadpole of *Didemnum molle,* with large cloud of attached *Prochloron* cells. White arrow: Adhesion disc at anterior end of larva. Black arrow: Eyespot. About 20 X.

Figures 6–9. Light microscopy and nucleic acid distribution.

Figure 6. Whole cells removed from *L. patella* and fixed in glutaraldehyde, unstained. 1200 X.

Figure 7. Whole cells from *Didemnum molle,* fixed in glutaraldehyde, unstained. These cells lack the large central vacuole of *Prochloron* from *L. patella.* They also average somewhat larger, and most contain a characteristic brown pigment. 1,200 X.

Figure 8. RNA distribution of *Prochloron* from *L. patella.* Branchial stigmata containing cells of *Prochloron.* Material was fixed in glutaraldehyde, embedded in glycol methacrylate and sectioned at 2mm. Staining was in azure B (0.1% in citrate buffer pH 4.0). Staining of ribosomal RNA is evident in the stromal areas, arranged as concentric lamellae around the central vacuole. The cell walls also stain, probably because of acid polysaccharides. Note the absence of stain in the medial sulcus of the cell wall (arrows). The small densely staining particles lining the stigmata are bacteria. About 600 X.

Figure 9. DNA distribution in cells expressed from *L. patella,* fixed in glutaraldehyde, embedded in glycol methacrylate and sectioned at 2 μm. Sections were digested with ribonuclease (0.1 mg per ml, at 37°C) for 1 hr and stained for DNA with DAPI (10 λg per ml in phosphate buffer pH 7.4). DNA containing regions form a network in the areas of stroma. The bright round objects are host cell nuclei. About 600 X. See also Color Figure 29.

Figure 10. *Prochloron* from *L. patella,* showing the typical nearly spherical shape and large, usually empty-appearing, central vacuole. Major cell structures include extensive areas of stacked thylakoids (t), interspersed with concentric regions of stroma (cytoplasm) (s). Stromal areas contain carboxysomes (C), occasional large crystalloids (x) and dense granules, probably polyphosphate (p). 10,000 X. Bar equals 1 μm.

Figure 11. Detail of stroma region, showing DNA filaments (arrows) and clusters of ribosomes (r). The dark structure at the right is a carboxysome (c). 60,000 X. Bar equals 1 μm.

Figures 12–14. Variation in morphology of *Prochloron* cells from different didemnid hosts. All about 10,000 X. Bar equals 1 μm.

Figure 12. *Prochloron* from *Diplosoma virens.* Stromal areas are prominent and occupy much larger area than thylakoids. Most thylakoids are only paired.

Figure 13. *Prochloron* from *Lissoclinum voeltzkowi,* with large uniformly dense granular areas in stroma and central vacuole. Carboxysome (c).

Figure 14. *Prochloron* from *Didemnum molle,* showing numerous small vacuoles, but no large central vacuole, numerous osmiophilic granules, probably the location of the characteristic brown pigment. Thylakoids (t) are arranged concentrically near the cell wall, but are irregularly arranged throughout the cell Carboxysome (c). Compare with Figures 28, 36 and 37.

Figures 15 and 16. Cells of the cyanobacterium *Anabaena*, to demonstrate the marked difference in thylakoid structure between *Prochloron* and bluegreen algae. Cells have been fixed in glutaraldehyde and post-fixed with 1 potassium permanganate (preparation by Kristin Black). Thylakoids are widely and evenly separated, associated with the presence of phycobilisomes. Carboxysomes (c) occur in stromal areas. The prominent granules in Figure 16 are glycogen-like storage products. Figures 15, about 50,000 X. Bar equals 1 μm. Figure 16, about 80,000 X. Bar equals 0.1 μm.

Figures 17–20. Cell division.

Figure 17. Cell from *L. patella* in midstage of division. Carboxysomes (c). About 4,500 X.

Figures 18 and 19. Beginnings of crosswall formation in *L. voeltzkowi*. The forming partition is surrounded by a margin of stroma, similar to the stromal region that borders the cell membrane. Thylakoids are indented into the central vacuole as the crosswall enlarges. About 30,000 X.

Figure 20. A completed crosswall in cell from *L. patella*, bordered by a very narrow rim of stroma. 8,000 X.

Figures 21–23. *Prochloron* from *L. patella* fixed in glutaraldehyde, post-fixed with 1% potassium permanganate to demonstrate arrangement of thylakoid membranes. Stromal structures (ribosomes, carboxysomes) are not preserved.

Figure 21. 12,000 X. Membranes in the central vacuole are probably a fixation artifact.

Figures 22 and 23. 36,000 X. Thylakoids are frequently arranged in stacks of 2 or 3. Unpaired thylakoids are rarely seen.

Figures 24–27: Freeze-fracture preparations of *Prochloron* cells from *D. virens*. From Giddings, Withers and Staehelin (1980) with kind permission of the authors.

Figure 24. Thin section of *Prochloron* cell. 7,500 X.

Figure 25. Freeze-fracture replica of *Prochloron* cell fixed in glutaraldehyde, frozen in glycerol-seawater in liquid Freon, then transferred to liquid nitrogen, fractured and the surface replica made with a Balzers freeze-etch apparatus (11,000 X).

Figure 26. Freeze-fracture micrograph showing appression of adjacent thylakoid membranes. Areas of membrane contact (shown by arrows) show a greater density of particles on the E fracture face. Stomal area (c) and thylakoid lumen (asterisk) are also evident (51,000 X).

Figure 27. This preparation was frozen in 70% glycerol without prior fixation. Stacked and unstacked regions are shown. A high ridge (between arrows) denotes a region where membranes are separated. When the ridge is small, the membranes are in contact. A marked difference in particle density is evident on the E face when stacked (EFs) and unstacked (EFu) are compared (76,000 X).

Figures 28–34. Components of *Prochloron* cell walls.

Figure 28. Cell surface of *Prochloron* from *Didemnum molle*, showing extracellular matrix of fine filaments and club-shaped processes arising from a relatively thin two-layered cell wall. The underlying stroma contains a small crystalloid (x) and clusters of inclusion granules (32,000 X. Bar represents 1 μm).

Figure 29. Cell wall of *Prochloron* from *L. patella*, showing typical multilayered structure. A rim of dense stromal material underlies the cell membrane. Outside the cell membrane is an area of low electron density, then a dense layer, a second area of low density, and our outer dense layer, from which matrix filaments arise. The inner low density layer appears in places to be divided by a thin dense line. (90,000 X. Bar represents 9.1 μm.)

Figure 30. Scanning electron micrograph of cell from *L. patella* that has ruptured at the medial sulens. This is a region where the cell wall appears to be thinner and subject to rupture in some cells during fixation (cf. Figure 8). Surfaces of cells from *L. patella* appeared to be nearly smooth at this magnification. About 7,000 X. Bar equals 1 μm.

Figure 31. Scanning electron micrograph of dividing cell from *Didemnum molle*. These cells all appeared somewhat roughened on the outer surface. About 8,000 X. Bar equals 1 μm.

Figure 32. Scanning micrograph of dividing cell from *L. patella* that also has ruptured along the sulcus, demonstrating that the sulcus and division plane occur perpendicular to one another. About 3,000 X. Bar equals 1 μm.

Figure 33 and 34. Phase micrographs of cell walls of *Prochloron* cells freshly expressed from *L. patella* into hypotonic sea water. The cells have ruptured from osmotic shock, and contracted into clam-shell shaped structures. About 1,200 X.

Figures 35–37. Some common inclusions in *Prochloron.*

Figure 35. Two inclusions of cells from *L. patella.* A portion of a large protein-containing crystal is shown with a complex structure produced by Moire effects. The large spacing is about 11 nm, but the thinner region at the upper left contains a 7.5 nm spacing. Most cells contain at least one crystalloid of this type (cf. Figure 10), at least some of which are colored bright orange red, probably from their carotenoid content. The smaller darker crystalloid with a parallelogram shape at lower left is frequently associated with the larger crystalloids. 76,000 X. Bar equals 0.1 μm.

Figure 36. A small crystalloid in cells from *Didemnum molle* with a spacing of about 9.2 nm (see also Figure 28). Two kinds of non-crystalloid inclusions are also shown, one dense and polygonal, the other less dense and round in outline. 45,000 X. Bar equals 0.1 μm.

Figure 37. A carboxysome, labeled with antibodies against RUBISCO, tagged with colloidal gold particles. The tissue was fixed only in glutaraldehyde, with no osmium tetroxide. The two types of inclusions shown in Figure 36 are not electron dense in this preparation, thus their density following osmium tetroxide probably indicates their lipid content. (See Swift and Leser, 1989.) 30,000 X. Bar equals 1 μm.

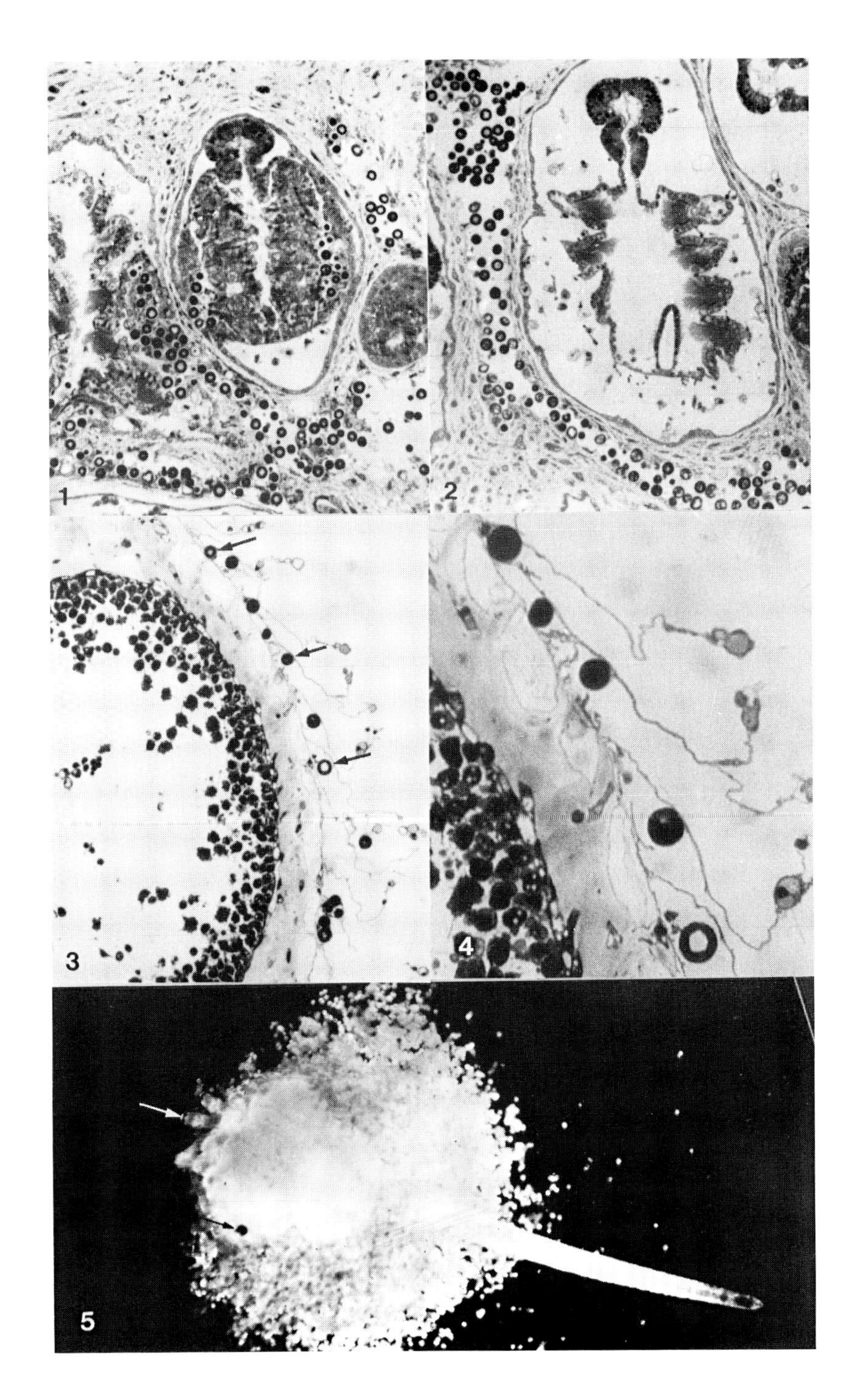
1
2
3
4
5

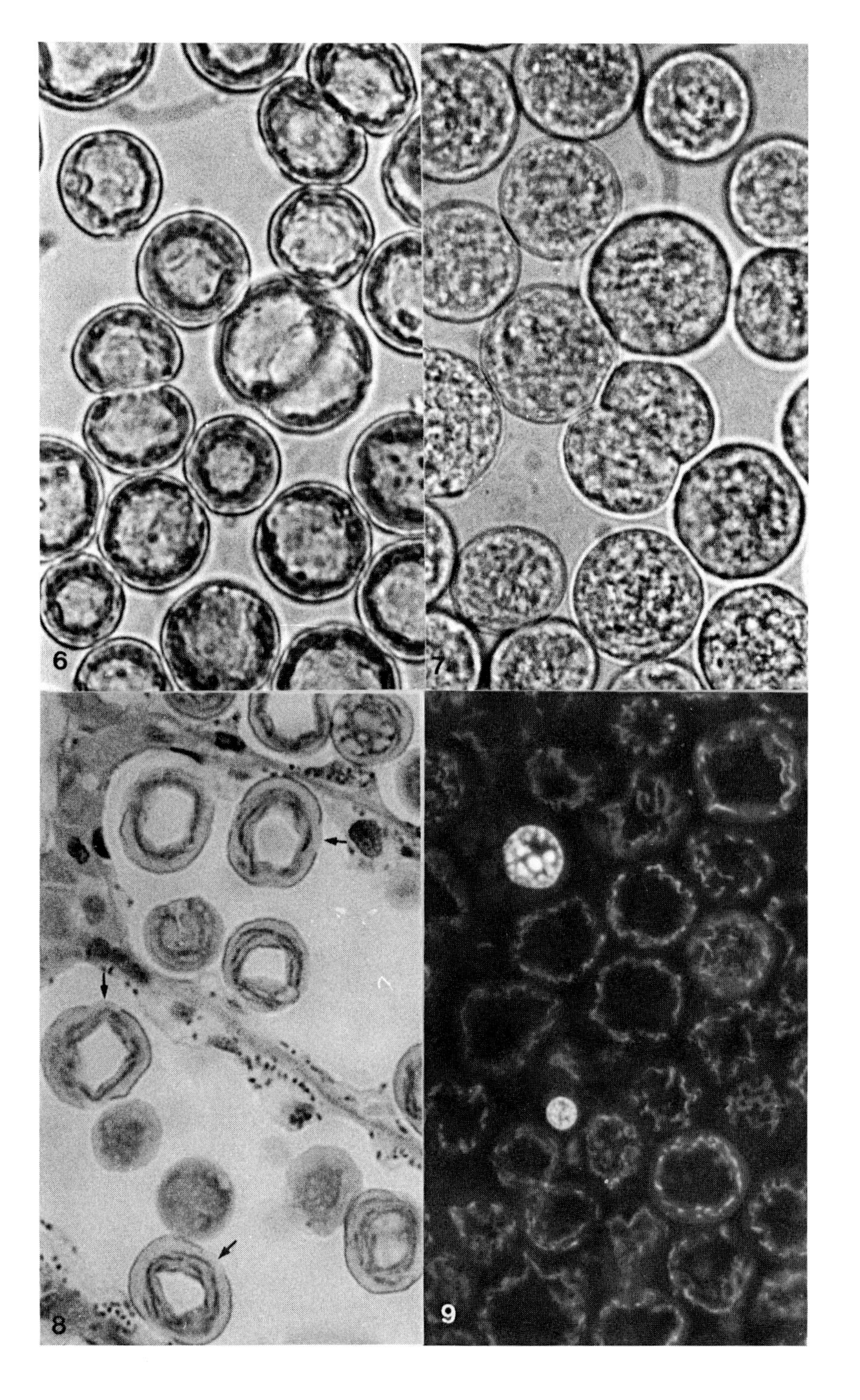
6
7
8
9

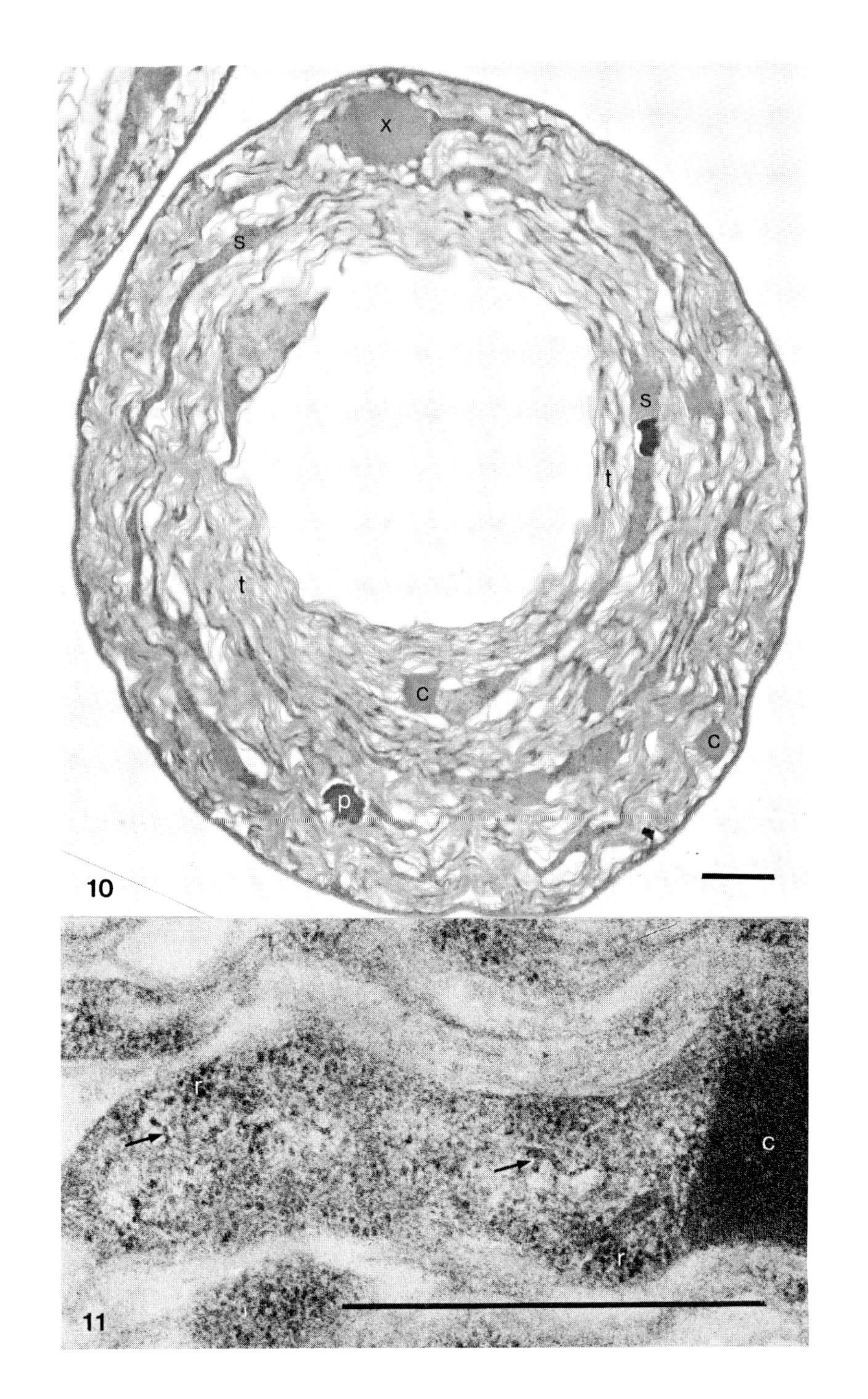
x
s
s
t
t
c
c
p
10
r
c
r
11

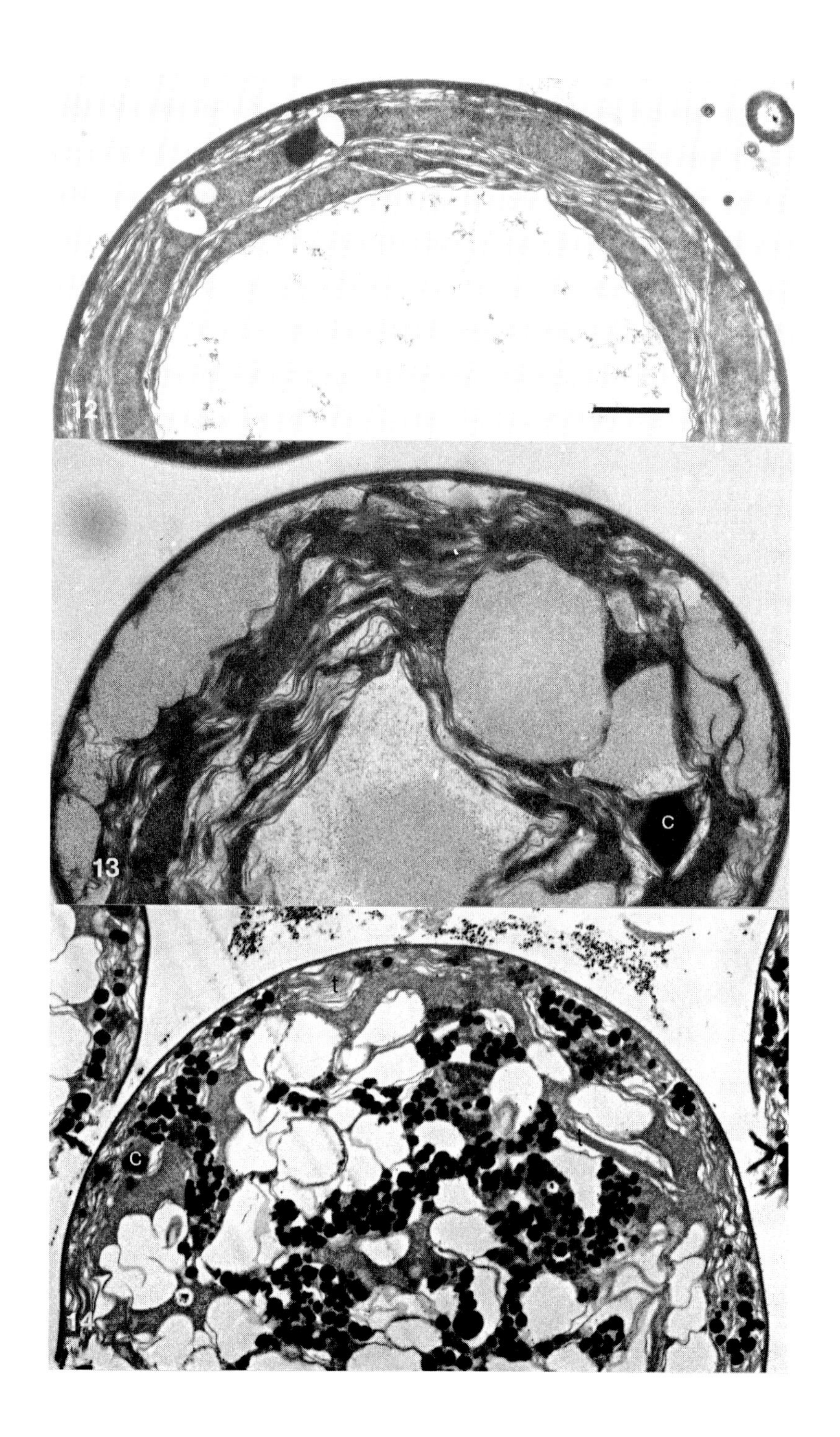
12
13
c
14
c
t
t

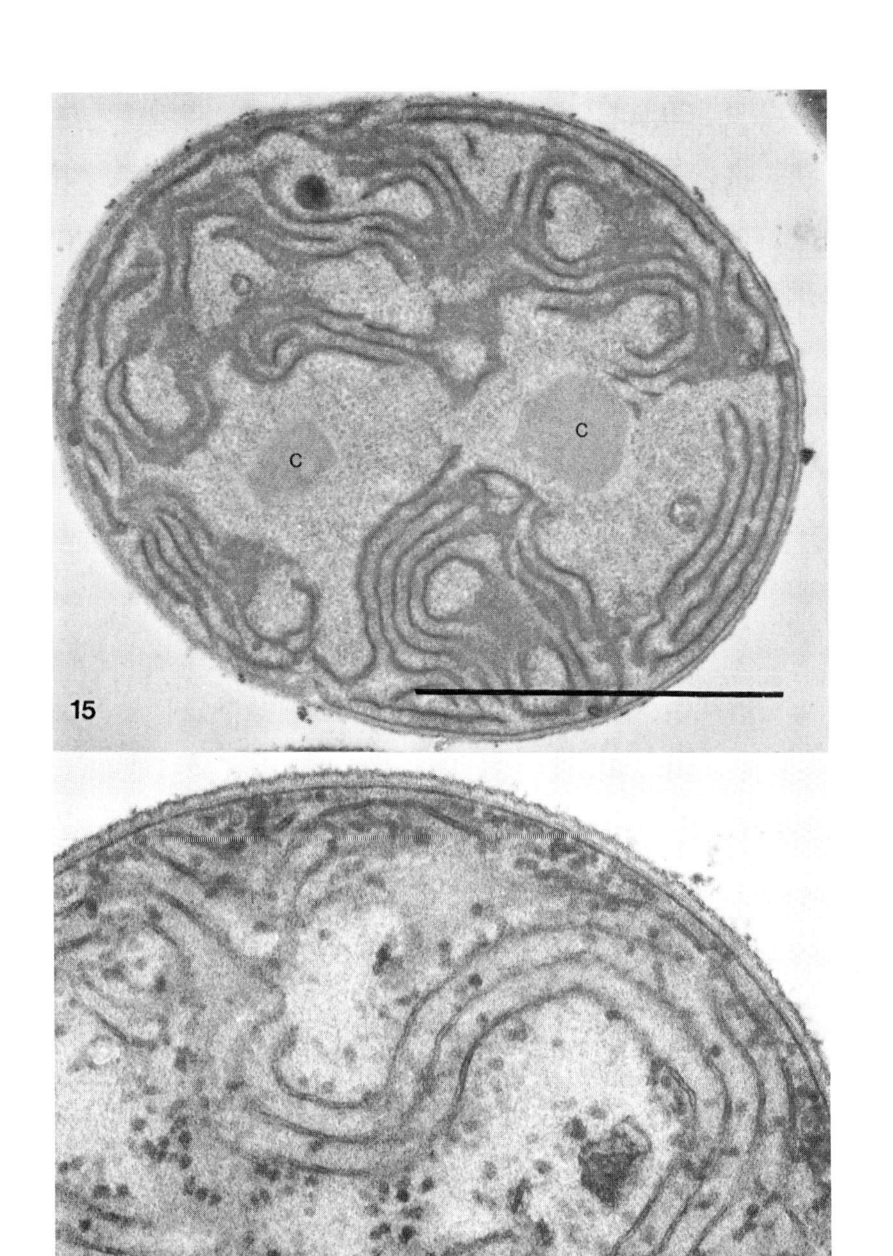
C
C
15
16

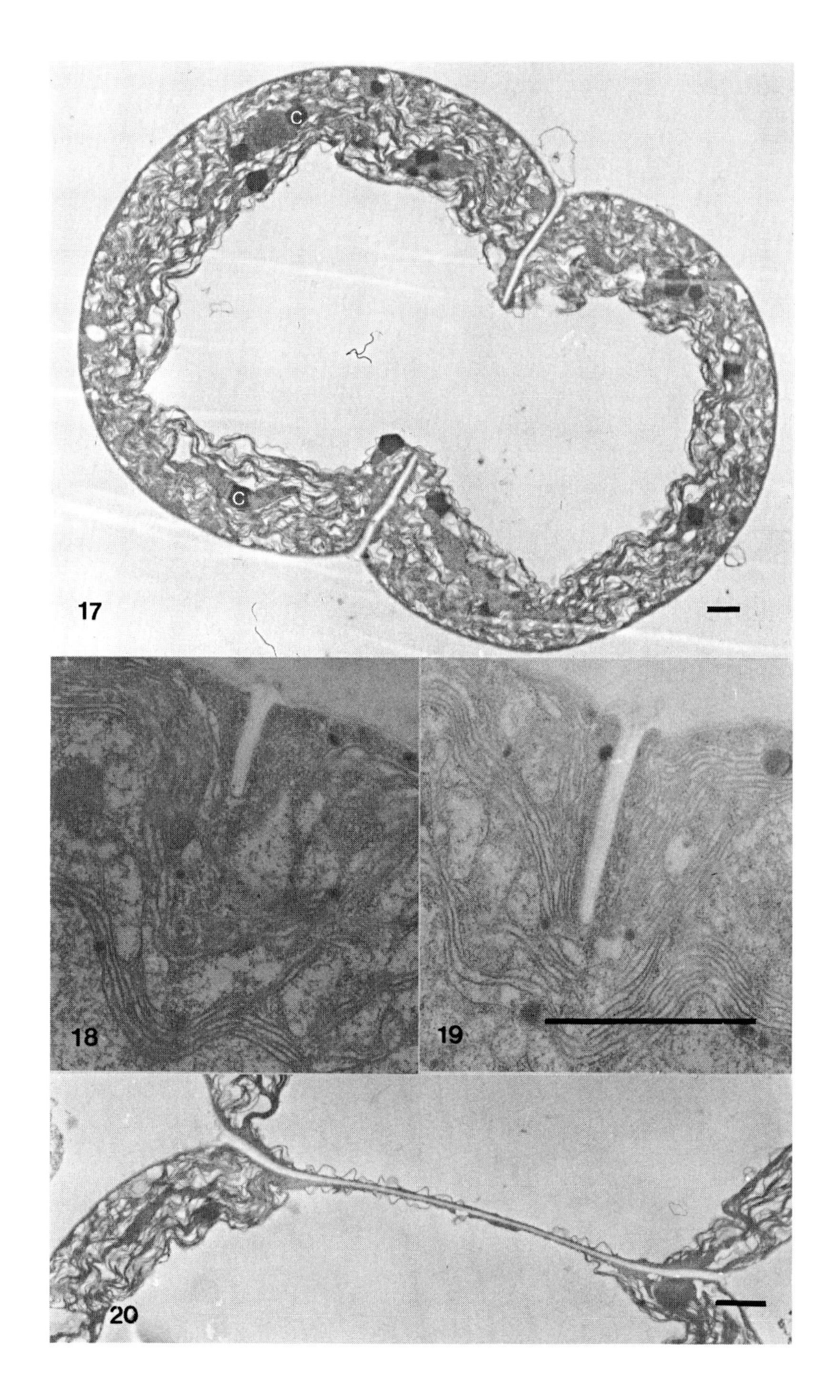

C
C
17
18
19
20

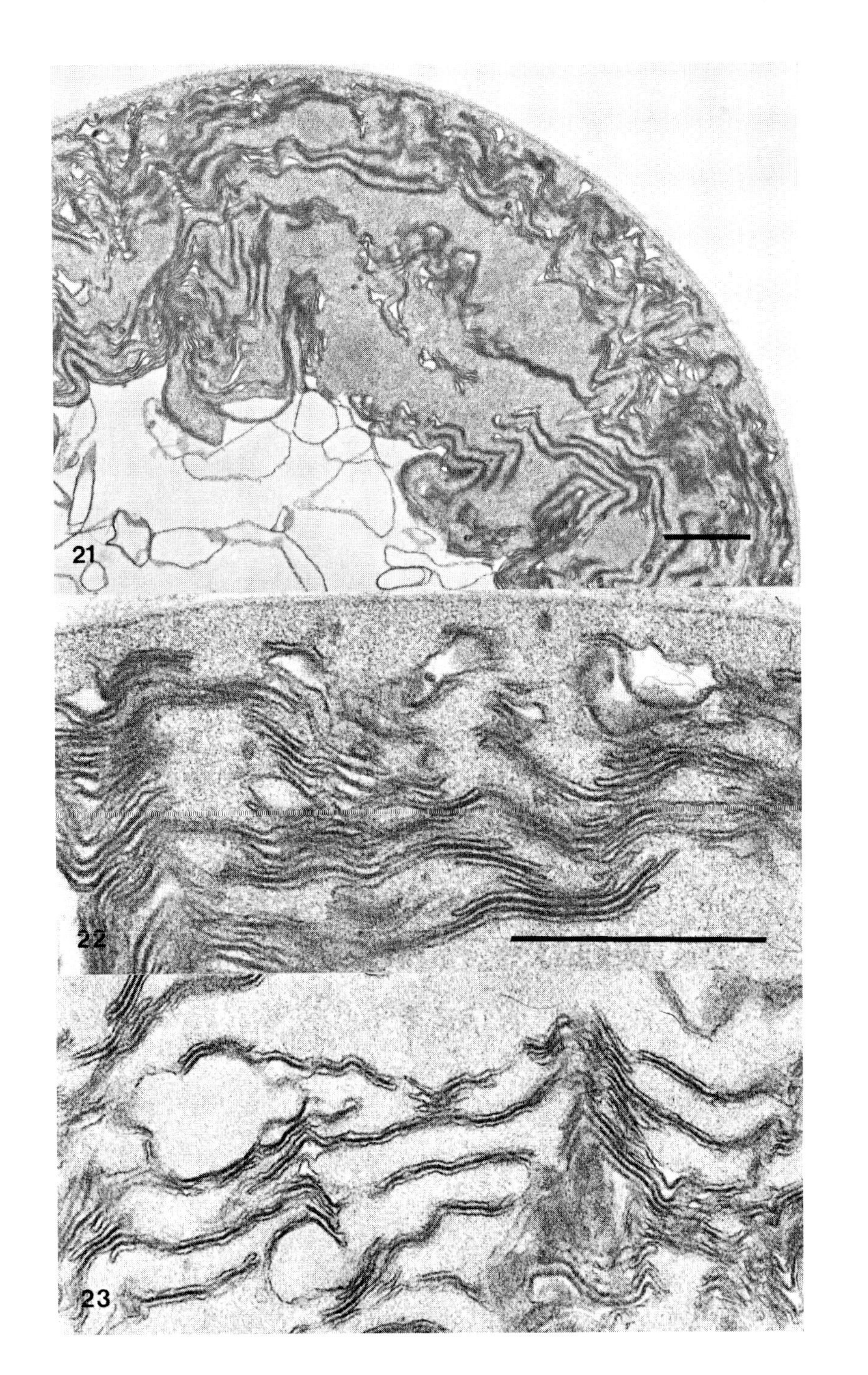
21
22
23

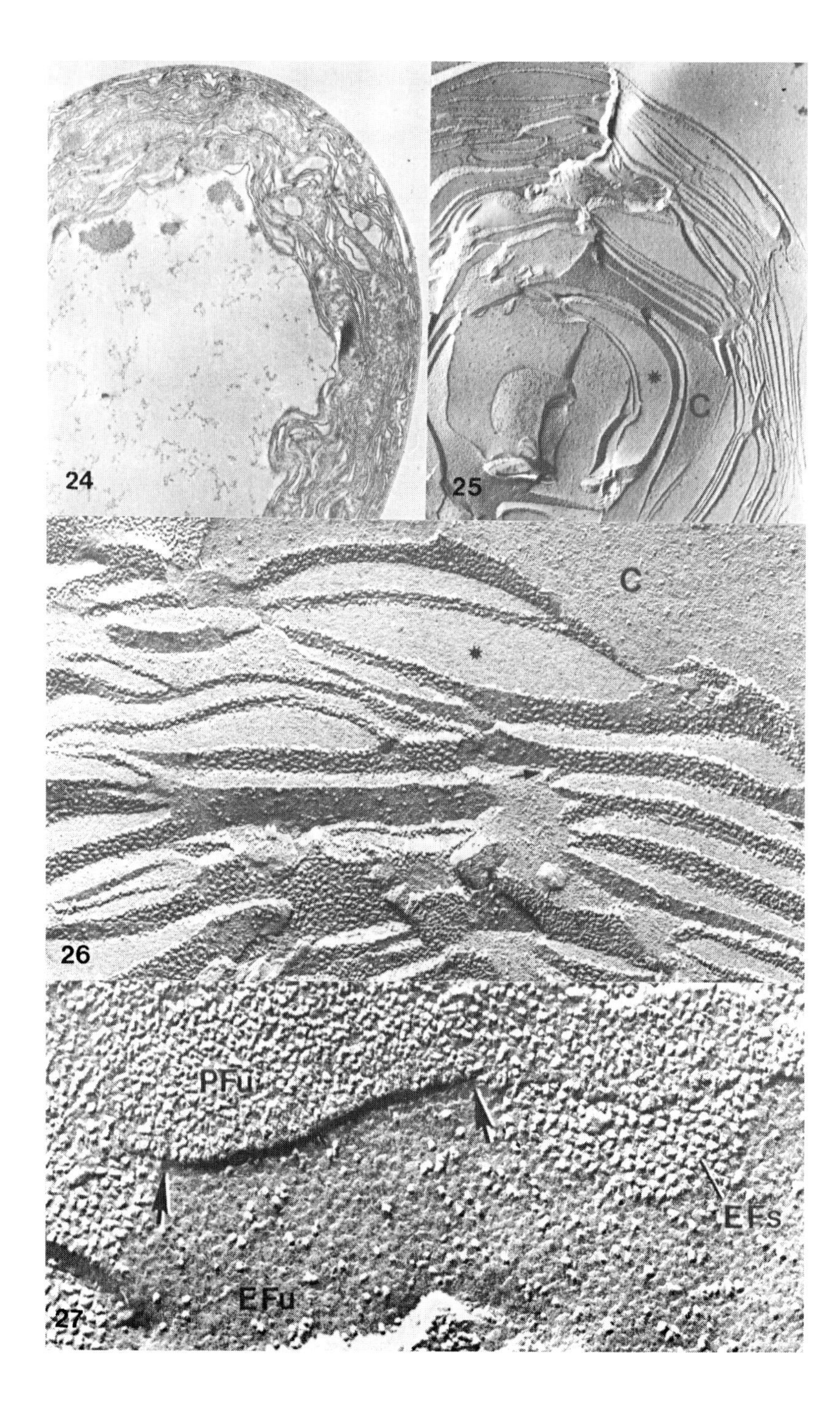
24
25
C
26
C
PFu
EFs
EFu
27

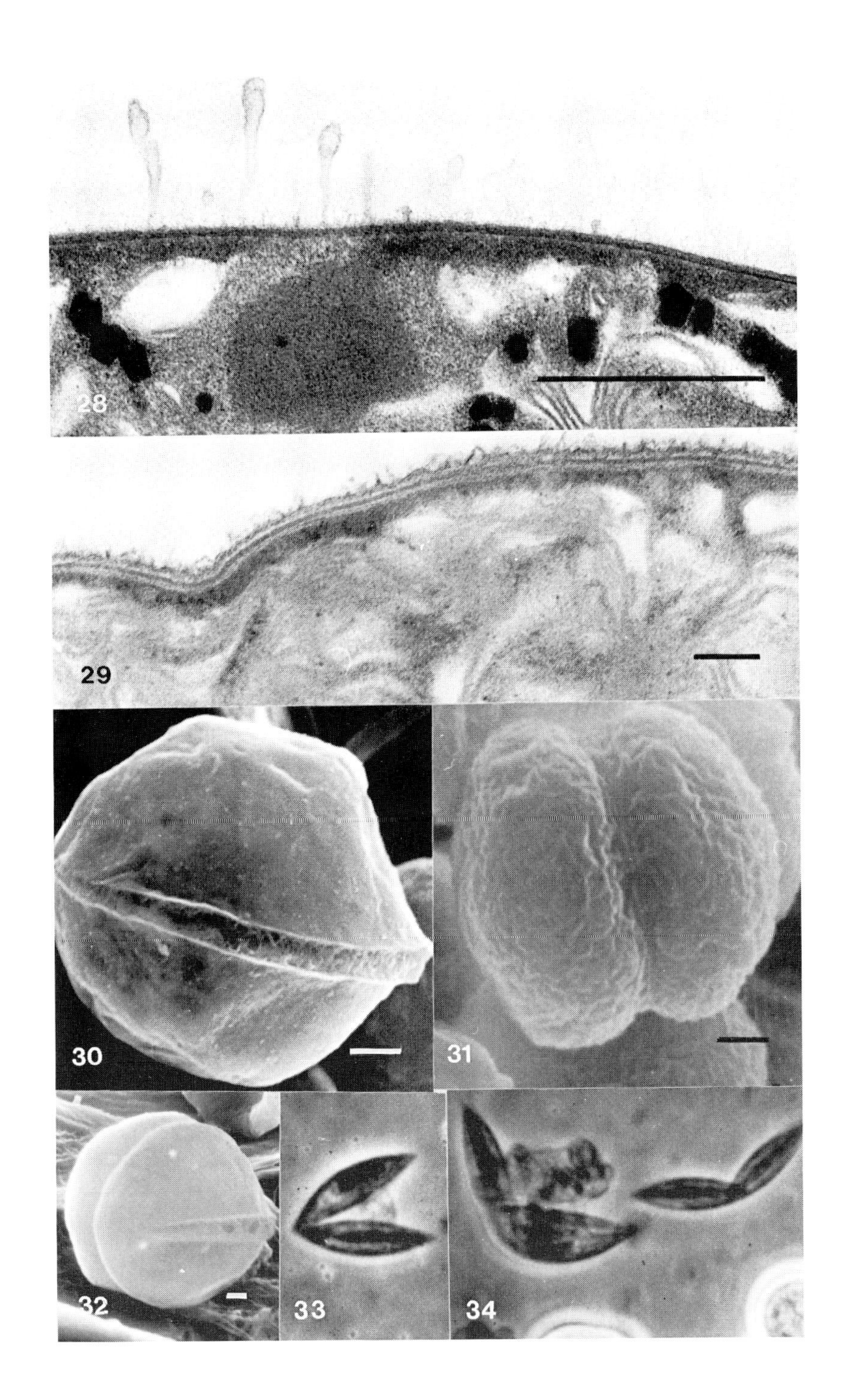
28
29
30
31
32
33
34

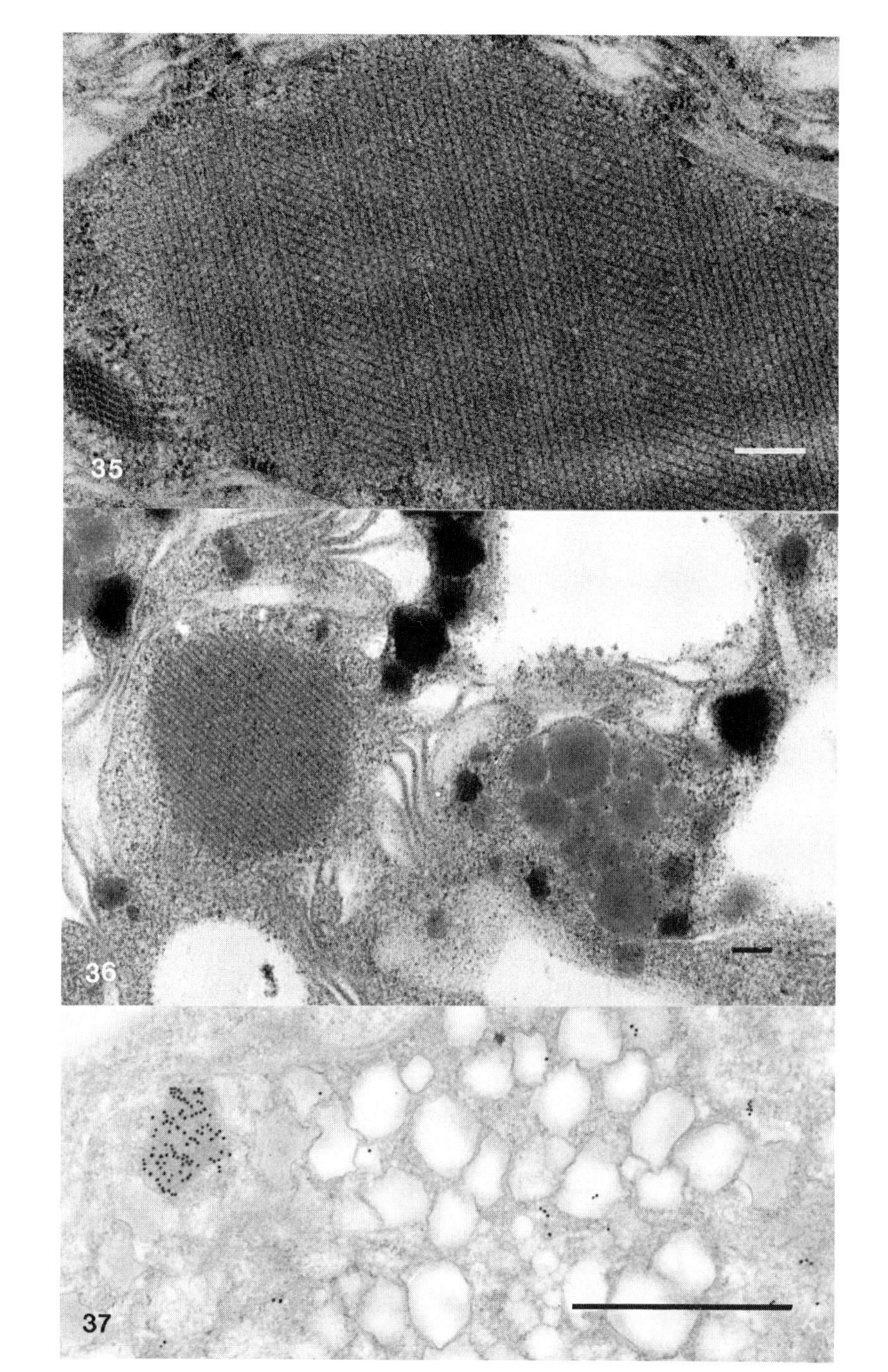
35
36
37

Chapter 8

Status Report on *Prochlorothrix hollandica*, a Free-Living Prochlorophyte

Hans C. P. Matthijs, Tineke Burger-Wiersma,*
and Luuc R. Mur

Ever since the description of *Prochloron* sp. (Lewin and Withers, 1975; Lewin, 1976, 1977), the search for a nonsymbiotic species with analogous properties became obvious. Such an organism was discovered in the Loosdrecht Lakes (The Netherlands) (Burger-Wiersma et al., 1986), a shallow freshwater lake system that originated from peat excavation.

Because of its shallowness (2 m) the lake is well mixed through the entire water column throughout the year. During spring and late fall the phytoplankton is dominated by cyanobacteria. In summer, when biomass reaches its highest levels (100–150 g Chl al^{-1}), the phytoplankton is dominated by a cyanobacterium-like species somewhat resembling *Oscillatoria limnetica*. In connection with a lake restoration project it was

The authors are indebted to Drs. G. S. Bullerjahn and L. Sherman (University of Missouri at Columbia; USA), Dr. Gy. Garab (National Academy of Sciences; Szeged, Hungary), Dr. K. R. Miller (Brown University; Providence, RI, USA), and Dr. L. J. Stal (University of Oldenburg, FRG) for making results available prior to publication. We also thank Dr. R. S. Alberte (University of Chicago) for critical comments.

T. B-W. is financially supported by the Dutch Ministry of Public Housing, Physical Planning and the Environment.

H. C. P. M. was a recipient of a postdoctoral fellowship of the Netherlands Organization for the Advancement of Pure Research ZWO (grant urg. 83-84).

isolated for ecophysiological studies. The cultures were yellowish green. Pigment analysis by HPLC revealed that the organism contained chlorophyll *b* as well as chlorophyll *a*, a combination hitherto found only in green-plant chloroplasts and *Prochloron*. Phycobilisomes and phycobilin pigments were absent. The carotenoid composition resembled that of cyanobacteria since no α-carotene or lutein was detected by HPLC, whereas β-carotene and zeaxanthin were major constituents. The prokaryotic nature of the cells was established by transmission electron microscopy (TEM) (Burger-Wiersma et al., 1986). The presence of chlorophyll *b* in addition to *a*, on the one hand, and the cyanobacterium-like cell structure and carotenoid composition on the other hand, indicated a possible relationship to *Prochloron*.

However, as Figure 8.1 shows, the morphology is completely different from that of *Prochloron*. Our organism forms trichomes of long cylinder-shaped cells, whereas *Prochloron* is a spherical, unicellular organism. Table 8.1 summarizes the similarities and differences between *Prochloron* and this new prochlorophyte for which we proposed the name *Prochlorothrix hollandica* (Burger-Wiersma, Stal, and Mur, in press). A summary of published and unpublished data follows. In contrast to *Prochloron*, *Prochlorothrix* can easily be grown in suspension culture on mineral media such as FPG (Burger-Wiersma, Stal, and Mur, submitted) or BG 11 (Allen, 1968). It remains viable on agar. *Prochlorothrix* can be grown at

Table 8.1 Comparison of some properties of *Prochloron* sp. (from the literature) and *Prochlorothrix hollandica*.

	Prochloron sp.	*Prochlorothrix hollandica*
Habitat	marine symbiotic	freshwater free-living
Morphology	prokaryotic unicellular cells spherical 10–20 μm in diam.	prokaryotic multicellular, filamentous cells cylindrical 1 μm wide
Pigments	Chl *a* and *b* *a*/*b* ratio 4–8 no phycobiliproteins no α-carotene no lutein echinenone	Chl *a* and *b* *a*/*b* ratio >7.5 no phycobiliproteins no α-carotene no lutein no echinenone
Physiology	host needed N_2-fixation* >25°C	mineral medium suffices no N_2-fixation 20°–30°C

*Recorded in one host species only.

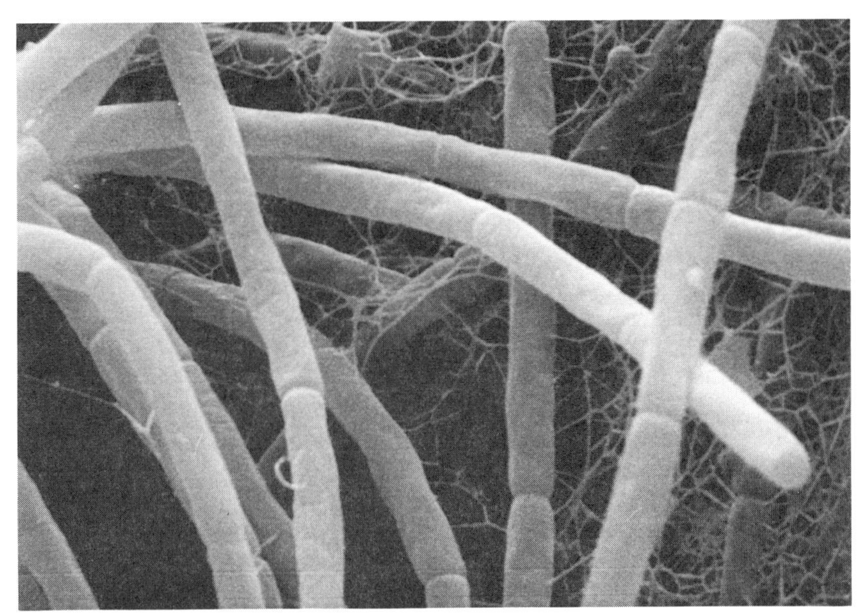

Figure 8.1 Scanning electron micrograph of *Prochlorothrix hollandica* (by Mrs Sabine Seufer, University of Oldenburg, FRG).

a variety of light intensities ranging from 4 to 400 $Em^{-2}s^{-1}$ (Burger-Wiersma, unpublished results). The strain is intermediate between cyanobacteria and green algae with respect to light–shade adaptation (Burger-Wiersma and Post, submitted).

Light-state transitions suggest the involvement of a light-harvesting chlorophyll *a/b* protein, which could be functionally similar to that known in green plants (Burger-Wiersma and Post, submitted). Redox-controlled reversible phosphorylation of a 29-kD protein of *Prochlorothrix* plays a role in the regulation of the light-energy distribution (Van der Staay *et al.*, in press).

The LHC complex in the thylakoid membranes of chloroplasts has been linked to stacking and has been assigned a role in the regulation of the distribution of light energy toward photosystem I and II by reversible phosphorylation of the apoprotein (Bennett, 1979). Stacking is not very prominent in TEM micrographs of *Prochlorothrix* (Burger-Wiersma et al., 1986). However, the associated lateral separation of photosystem I and II has been revealed by freeze-etch electron microscopy studies (Miller, 1987; Miller, Jacob, Burger-Wiersma, and Matthijs, 1988).

In non-denaturing polyacrylamide-gel electrophoresis of isolated thylakoid membranes of *Prochlorothrix* chlorophyll *b* appears to be bound to a Chl *a/b* protein complex (Bullerjahn et al., 1987). The estimated apparent molecular mass of the complex in SDS-PAGE is 34 kD. Antibodies raised against the denatured LHC of various higher plants and green algal chloroplasts do not show immunological cross-reactivity toward the chlorophyll *a/b* protein complex of *Prochlorothrix* (Bullerjahn et al., 1987; Fawley and Matthijs, unpublished results). Identical estimates of the molecular mass and immunological responses have been reported for *Prochloron* (Hiller and Larkum, 1985). Although the range of molecular mass estimates of green plant LHCII is 24–29 kD (Murphy, 1986), the estimate of 34 kD for prochlorophyte LHCII seems not to fit into this range. However, the homology of the green-plant light-harvesting complex (LHCII) to the chlorophyll *a/b* complex of *Prochloron* was indicated by their similar molecular masses (Withers et al., 1978). Comparisons of the native protein and its subunit composition and molecular sizes remain to be examined before we can draw definite conclusions on the phylogenetic nature of the chlorophyll *a/b* complex in prochlorophytes.

Prochlorothrix has a relatively high chlorophyll $a{:}b$ ratio of >7.5 (Table 8.1). The question of whether this relative deficiency in chlorophyll *b* content is reflected in the molecular structure of the chlorophyll *a/b* complex has been studied by spectroscopic analysis of the complex after its isolation on a nondenaturing gel, on which a lower ratio of approximately 4 was routinely found. In marked contrast to this, the ratio in chloroplast LHC is approximately 1.2 (Murphy, 1986). Circular dichroism features suggest that chlorophyll *b* is present in a monomeric form, and its orientation is different from that known for LHCII isolated from higher-plant chloroplasts (Matthijs *et al.*, in press).

The results presented earlier demonstrate that the chlorophyll *a/b* complex of *Prochlorothrix* may be somewhat different from the LHC complexes of the higher plant chloroplast in molecular and structural composition, but may be functionally analogous.

An analysis of the lipids of *Prochlorothrix*, with their implications for the chemotaxonomy of this unusual organism, has been recently published (Volkman et al., 1988). Forthcoming research will aim at a detailed description of the photosynthetic processes in this organism with emphasis on biochemical, physiological, genetic, and ecological aspects, which will allow comparisons to other oxygenic photosynthetic organisms. The discovery of *Prochlorothrix hollandica* has evoked renewed discussions on the origin of chloroplasts (Cox, 1986; Walsby, 1986) and has stimulated interest in its possible usefulness in research (Barber, 1986; Miller, 1987).

A recently published phylogenetic tree (based on 16S rRNA sequences) indicates that *Prochlorothrix*, like *Prochloron*, is closer to cyanophytes than to chloroplasts (Turner et al., 1989), although another tree (based on amino-acid sequences in one of the thylakoid proteins of photosystem II) puts *Prochlorothrix* nearer to the chloroplast line (Morden and Golden,

1989). We are confident that such disagreements will be resolved in the near future.

References (Additional to Bibliography)

Allen, M. M. Simple conditions for growth of unicellular blue-green algae on plates. J. Phycol. 4:1–4; 1968.

Barber, J. New organism for elucidating the origin of higher plant chloroplasts. Trends Biochem. Sci. 11:238; 1986.

Bennett, J. Chloroplast phosphoproteins. Dephosphorylation of polypeptides of the light-harvesting chlorophyll protein complex. Eur. J. Biochem. 99:133–137; 1979.

Bullerjahn, G. S., H. C. P. Matthijs, L. R. Mur, and L. Sherman. Chlorophyll-protein composition of the thylakoid membrane from *Prochlorothrix hollandica*, a prokaryote containing chlorophyll *b*. Eur. J. Biochem. 168:295–300; 1987.

Burger-Wiersma, T., M. Veenhuis, H. J. Korthals, C. C. M. Van de Wiel, and L. R. Mur. A new prokaryote containing chlorophylls *a* and *b*. Nature 320:262–264; 1986.

Burger-Wiersma, T., L. J. Stal, and L. R. Mur. *Prochlorothrix hollandica* gen. nov., spec. nov.: a filamentous, oxygenic photoautotrophic prokaryote containing chlorophylls *a* and *b*. *Int. J. Syst. Bacteriol;* 1989.

Cox, G. The origin of chloroplasts in eukaryotes. Nature 322:412; 1986.

Hiller, R. G., and A. W. D. Larkum. The chlorophyll-protein complexes of *Prochloron* sp. (Prochlorophyta). Biochim. Biophys. Acta 806:107–115; 1985.

Matthijs, H. C. P., G. W. M. van der Staay, H. van Amerongen, R. van Grondelle, and G. Garab. Structural organization of chlorophyll *b* in the prochlorophyte *Prochlorothrix hollandica*. Biochem. Biophys. Acta, In press.

Miller, K. R. Studies on photosynthetic membrane organization. UCLA Symposium on Molecular and Cellular Biology, n.s., 63:27–46; 1987.

Miller, K. R., J. S. Jacob, T. Burger-Wiersma, and H. C. P. Matthijs. Supramolecular structure of the thylakoid membrane of *Prochlorothrix hollandica:* a chlorophyll *b*-containing prokaryote. J. Cell Sci. 91:577–586; 1988.

Morden, C. W., and S. S. Golden. psbA genes indicate common ancestry of prochlorophytes and chloroplasts. Nature 337:382–385; 1989.

Murphy, D. J. The molecular organization of the photosynthetic membranes of higher plants. Biochim. Biophys. Acta 864:33–94; 1986.

Turner, S., T. Burger-Wiersma, S. J. Giovannoni, L. R. Mur, and N. R. Pace. The relationship of a prochlorophyte *Prochlorothrix hollandica* to green chloroplasts. Nature 337:380–382; 1989.

Van der Staay, G. W. M., H. C. P. Matthijs, and L. R. Mur. Phosphorylation and dephosphorylation of membrane proteins from the prochlorophyte *Prochlorothrix hollandica* in fixed redox states. Biochem. Biophys. Acta, in press.

Volkman, J. K., T. Burger-Wiersma, P. D. Nichols, and R. E. Summons. Lipids and chemotaxonomy of *Prochlorothrix hollandica*, a planktonic prokaryote containing chlorophylls *a* and *b*. *J. Phycol.* 24:554–559; 1988.

Walsby, A. E. Origins of chloroplasts. Nature 320:212; 1986.

Epilogue

*The Emerald Crystal Ball**

When the tides are warm and low
Where the tropic sun has shone,
That is where we look for Pro-
chloron.

On the shores of Mexico,
Eniwetok and Ceylon
Lurk didemnids bearing Pro-
chloron.

There are things we ought to know—
Mysteries to think upon—
Problems that relate to Pro-
chloron.

How to get the cells to grow:
Media to grow them on—
These are what we need for Pro-
chloron.

Progress has been somewhat slow
Towards our chosen Rubicon:
How to tame the tiny Pro-
chloron.

* * *

Prince, if you desire to know
Where the last year's snows have gone
Peer into the heart of Pro-
chloron.

*From "The Biology of Algae and Diverse Other Verses," by Ralph A. Lewin, The Boxwood Press, 1987, p. 13.

Bibliography

Published Articles on *Prochloron* (with Abstracts)

Akazawa, T., E. H. Newcomb, and C. B. Osmond. Pathway and products of Co$_2$-fixation by green prokaryotic algae in the cloacal cavity of *Diplosoma virens.* Mar. Biol. 47:325–330; 1978.

Light-dependent $^{14}CO_2$ fixation by the algae of *Diplosoma virens* (Hartmeyer) ranged between about 3 and 27 μmoles mg^{-1} chlorophyll h^{-1}. The principal first products of ^{14}C fixation were 3-phosphoglyceric acid and phosphorylated sugars, indicating that ribulose bisphosphate carboxylase was the primary carboxylation enzyme. The activity of this enzyme in crude extracts of the algae was 4 to 6 μmoles CO_2 mg^{-1} chlorophyll h^{-1}. The principal end product of ^{14}C fixation by these algae in the ascidian host was a water-soluble oligosaccharide which was an α–1,4-glucan. A maximum of 7% of the ^{14}C fixed was found in insoluble materials of the algae or its host after 60 min $^{14}CO_2$ fixation. Whether the α–1, 4-glucan is a product of algal or animal metabolism remains to be determined.

Alberte, R. S., L. Cheng, and R. A. Lewin. Photosynthetic characteristics of *Prochloron* sp./ascidian symbioses. I. Light and temperature responses of the algal symbiont of *Lissoclinum patella*. Mar. Biol. 90:575–587; 1986.

The prokaryotic green alga *Prochloron* sp. (Prochlorophyta) is found in symbiotic association with colonial didemnid ascidians that inhabit warm tropical waters in a broad range of light environments. We sought to determine the light-adaptation features of this alga in relation to the natural light environments in which the symbioses are found, and to characterize the temperature sensitivity of photosynthesis and respiration of *Prochloron* sp. in order to assess its physiological role in the productivity and distribution of the symbiosis. Colonies of the host ascidian *Lissoclinum patella* were collected from exposed and shaded habitats in a shallow lagoon in Palau, West Caroline Islands, during February and March, 1983. Some colonies from the two light habitats were maintained under conditions of high light (2,200 μE m^{-2} s^{-1}) and low light (400 μE m^{-2} s^{-1}) in running seawater tanks. The environments were characterized in terms of daily light quantum fluxes, daily periods of light-saturated photosynthesis (H_{sat}), and photon flux density levels. *Prochloron* sp. cells were isolated from the hosts and examined for their photosynthesis vs irradiance relationships, respiration, pigment content and photosynthetic unit features. In addition, daily P:R ratios, photosynthetic quotients, carbon balances and photosynthetic carbon release were also characterized. It was found that *Prochloron* sp. cells from low-light colonies possessed lower chlorophyll *a/b* ratios, larger photosynthetic units sizes based on both reaction I and reaction II, similar numbers of reaction center I and reaction center II per cell, lower respiration levels, and lower P_{max} values than cells from high-light colonies. Cells isolated from low-light colonies showed photoinhibition of P_{max} at photon flux densities above 800 μE m^{-2} s^{-1}. However, because the host tissue attenuates about 60 to 80% of the incident irradiance, it is unlikely that these cells are normally photoinhibited *in hospite*. Collectively, the light-adaptation features of *Prochloron* sp. were more similar to those of eukaryotic algae and vascular plant chloroplasts than to those of cyanobacteria, and the responses were more sensitive to the daily flux of photosynthetic quantum than to photon flux density *per se*. Calculation of daily minimum carbon balances indicated that, though high-light cells had daily P:R ratios of 1.0 compared to 4.6 for low-light cells, the cells from the two different light environments showed nearly identical daily carbon gains. Cells isolated from high-light colonies released between 15 and 20% of their photosynthetically-fixed carbon, levels sufficient to be important in the nutrition of the host. Q_{10} responses of photosynthesis and respiration in *Prochloron* sp. cells exposed briefly (15–45 min) to temperatures between 15° and 45°C revealed a discontinuity in the photosynthetic response at the ambient growth temperatures. The photosynthetic rates were found to be more than twice as sensitive to temperatures below ambient (Q_{10} = 3.47) than to temperatures above ambient (Q_{10} = 1.47). The Q_{10} for respiration was constant (Q_{10} = 1.66) over the temperature range examined. It appears that the photosynthetic temperature sensitivity of *Prochloron* sp. may restrict its distribution to warmer tropical waters. The ecological implications of these findings are discussed in relation to published data on other symbiotic systems and free-living algae.

Alberte, R. S., L. Cheng, and R. A. Lewin. Characteristics of *Prochloron*/ascidian symbioses. II. Photosynthesis-irradiance relationships and carbon balance of associations from Palau, Micronesia. Symbiosis 4:147–170; 1987.

Photosynthesis-irradiance (P-I) relationships, respiration, and photosynthesis/respiration ratios (P:R) are compared for six species of *Prochloron*/ascidian symbioses in Palau, West Caroline Islands: *Lissoclinum patella, Lissoclinum punctatum, Lissoclinum voeltzkowi, Diplosoma similis, Diplosoma virens,* and *Trididemnum cyclops*. The colonies were collected in a shallow coral lagoon in fully exposed, high photo-flux environments and in shaded environments. For the *L. patella* symbiosis, we determined the Q_{10} values for photosynthesis and respiration between 15 and 45°C and the fractional contribution of *Prochloron* carbon production to ascidian respiration demand. The P-I relationships of the colonies demonstrated photo-adaptation features similar to that of the freshly isolated symbiont and reflected adaptation to the photosynthetic quantum fluxes inside the animal rather than total quantum fluxes incident on the colonies. All of the colonies except *L. voeltzkowi* showed daily $P_{net}:R_{colony}$ ratios greater than 1.0 when determined by using daily periods of saturating and compensating light. When the daily $P_{net}:R_{animal}$ ratios and the contribution of symbiont carbon to host respiration were determined for high- and low-light colonies of *L. patella,* the ratios were between 1.5 and 2.8, while the carbon contribution from the symbiont to the host was between 30–56%. The log transformation of photosynthetic rate vs. temperature relationships yielded a discontinuity of the ambient growth temperature (30°C); the Q_{10} below ambient was 3.52 while it was 1.62 above ambient temperatures. The Q_{10} for colony respiration was 1.97 over the entire temperature range (15–45°C). The estimated production rates of these symbioses were 13.1 $\mu g\ C\ dm^{-2}\ d^{-1}$. Collectively the results indicate that *Prochloron* can make significant contributions to the nutrition of its hosts, that *Prochloron*/ascidian symbioses make substantial contributions to benthic production in coral reef systems and that the low temperature sensitivity of symbiont photosynthesis probably restricts the distribution of the symbioses to warm water.

Andrews, T. J., D. M. Greenwood, and D. Yellowlees. Catalytically active hybrids formed *in vitro* between large and small subunits of different procaryotic ribulose bisphosphate carboxylases. Arch. Biochem. Biophys. 234:313–317; 1984.

Ribulose bisphosphate carboxylase from the procaryotic green alga, *Prochloron* (the symbiont of *Lissoclinum patellum*), has eight large and eight small subunits, and a low affinity of CO_2, similar to that of cyanobacterial carboxylases. The small subunits were progressively removed from this carboxylase and from that from the cyanobacterium, *Synechococcus* ACMM 323, by twice-repeated, mild-acid precipitation. This procedure produced large-subunit octamers, greatly depleted in small subunits, as well as isolated small subunits. Catalytic activity of the large-subunit preparations were reconstituted with both homologous and heterologous small subunits. The reassembled enzymes were catalytically competent in all cases. When fully saturated with small subunits, the hybrid enzymes were only about 20% less active than the homologously reconstituted enzymes. Heterologous reconstitution underscores the essential function of the small subunits in catalysis.

Antia, N. J. A critical appraisal of Lewin's Prochlorophyta. Br. Phycol. J. 12:271–276; 1977.

Lewin's proposal to create the Prochlorophyta as a new algal division to cover thylakoid-containing prokaryotes producing chlorophylls *a* and *b*, but no phycobiliprotein, is considered unjustifiable, because the criteria used appear to be overrated and insufficient to demarcate these prokaryotes from the Cyanophyta. A plea is made to relax the rigidity of the criteria traditionally followed in defining the Cyanophyta in order to avoid unnecessary proliferation of prokaryotic algal divisions.

Bachmann, M., A. Maidhof, H. C. Schröder, K. Pfeifer, E. M. Kurz, T. Rose, I. Müller, and W. E. G. Müller. *Prochloron* (Prochlorophyta): Biochemical contributions to the chlorophyll and RNA composition. Plant Cell Physiol. 26:1211–1222; 1985.

The prochlorophyte *Prochloron,* a symbiont of the colonial ascidian *Didemnum molle,* was collected in the Indian Ocean around Giravaru (Maldives) in depths between 1 and 40 m. The chlorophyll *a* to *b* ratio of the algal symbionts was higher in colonies living between 1 and 6 m, compared to that determined for *Prochloron* from a depth of 30 m. This property for chromatic adaptation in correlation with changes in the total content of chlorophyll is dependent upon environmental factors. The association between *Didemnum* and *Prochloron* is only a facultative symbiosis. The size of the colonies, growing near the water surface is large (up to 3 cm), and it gradually decreases to 0.2 cm in a depth of 30 m dim locations. At a depth of 40 m the tunicates do not contain the algal symbionts.

Applying quantitative preparative isolation and sensitive immunological as well as biochemical detection techniques we have no evidence for the existence of poly(A) stretches in RNA species from *Prochloron*. Moreover, we failed to detect both sn/scRNAs and their proteins, typically associated with them in RNP complexes from eukaryotes. From the data we suggest that mRNA synthesis proceeds in *Prochloron* in a way similar to prokaryotes.

Barclay, W. R., J. M. Kennish, V. M. Goodrich and R. Fall. High levels of phenolic compounds in *Prochloron* species. Phytochemistry 26(3):739–743; 1987.

Strains of the prokaryotic alga *Prochloron*, occurring internally in a variety of ascidian hosts in the South Pacific Ocean, were determined to have high intracellular concentrations of phenolic compounds, ranging from 1.8 to 7.1% of the cell dry weight. Only the externally occurring *Prochloron* from *Didemnum candidum*, a species found in the Gulf of California, exhibited low concentrations of phenolic compounds. Investigations with enzyme protectants and the lack of intracellular coagulation in *Prochloron* from *D. candidum* suggest that phenolic substances may be the cause of intracellular coagulation in other strains of *Prochloron*. This process has inhibited the extraction and study of enzymes from these unique algae.

Berhow, M. A. and B. A. McFadden. Ribulose 1,5-bisphosphate carboxylase and phosphoribulokinase in *Prochloron*. Planta 158:281–287; 1983.

Cell-free extracts of *Prochloron didemni* were assayed for ribulose 1,5-bisphosphate (RuBP) carboxylase (EC 4.1.1.39) and phosphoribulokinase (EC 2.7.1.19), two key enzymes in the reductive pentose-phosphate cycle. In an RuBP-dependent reaction, the production of two molecules of 3-phosphoglycerate per molecule of CO_2 fixed was shown. Phosphoribulokinase activity was demonstrated by the production of ADP from ribulose 5-phosphate (Ru5P) and ATP and by measurement of ATP-, Ru5P-dependent $^{14}CO_2$ fixation in the presence of excess spinach RuBP carboxylase. When *Prochloron* RuBP carboxylase was purified from cell-free extracts by isopycnic centrifugation in reoriented linear 0.2 to 0.8 M sucrose gradients, the enzyme sedimented to a position which corresponded to that for the 520,000-dalton spinach enzyme. After polyacrylamide gel electrophoresis (PAGE) of *Prochloron* enzyme, a major band of enzyme activity corresponded to that for the spinach enzyme. Considerably more additional carboxylase activity was found in a less mobile species than was the case for spinach RuBP carboxylase. Sodium dodecyl sulfate-PAGE of the *Prochloron* enzyme indicates that it is composed of both large (molecular weight, MW = 57,500) and small (MW = 18,800) subunits.

Björn, G. and L. O. Björn. *Prochloron*—blind end or missing link? Svensk Bot. Tidskr. 76:43–45; 1982.

Prochloron is a prokaryotic organism which, in contrast to blue-green algae, possesses chlorophyll *b* and grana as higher plants do. Close relatives of *Prochloron* could have given rise to chloroplasts of green algae by entering into a symbiotic relationship with a nonphotosynthetic organism.

Chadefaud, M. Sur la notion de prochlorophytes. Rev. Algol., N.S., 13:203–206; 1978.

It is suggested that the acquisition of the type of pigmentation found in the Chlorophyceae—with chlorophylls *a* and *b*—by *Prochloron* was the result of a "macro-mutation"—*eds.*

Chapman, D. J., and R. K. Trench. Prochlorophyceae: Introduction and bibliography. Selected Papers in Phycology II (J. R. Rosowski & B. C. Parker, eds.), Phycol. Soc. America, pp. 656–658; 1982.

With the outstanding exception of chlorophyll *b*, the features of *Prochloron* indicate that it is a "good" cyanobacterium, lacking many chemical and morphological characteristics of green algal chloroplasts.—*eds.*

Cheng, L., and R. A. Lewin. *Prochloron* on *Synaptula*. Bull. Mar. Sci. 35(1):95–98; 1984.

Prochloron is reported as occurring in sporadic patches on the surface of a holothurian, *Synaptula lamperti* Heding, found on sponges in shallow water in the Kamori Channel, Koror, Palau, W.C.I.—*eds.*

Coleman, A. W., and R. A. Lewin. The disposition of DNA in *Prochloron* (Prochlorophyta). Phycologia 22:209–212; 1983.

As revealed by staining with a fluorescent dye (DAPI), the DNA of *Prochloron* occurs as 15–50 irregularly shaped aggregates, up to 5 μm long, scattered among the thylakoids around the periphery of the cell.—*eds.*

Cox, G. Engulfment of *Prochloron* cells by cells of the ascidian, *Lissoclinum*. J. Mar. Biol. Assoc. U.K. 63:195–198; 1983.

Serial-section electron micrographs have been obtained of the phagocytosis of *Prochloron* cells by amoebocytes of the ascidian *Lissoclinum voeltzkowi* (Michaelsen). This may answer the hitherto puzzling question of how the host tunicate benefits from photosynthetic symbionts situated in an excurrent water flow.

Cox, G. Comparison of *Prochloron* from different hosts. I. Structural and ultrastructural characteristics. New Phytol. 104:429–445; 1986.

Prochloron cells from a variety of ascidian hosts and from sites throughout the tropical Pacific and Indian Oceans have been studied by optical and electron microscopy. With a few exceptions, they can be placed into three typological groups which correlate closely with the type of host-symbiont relationship. This grouping is also compatible with the limited biochemical taxonomic data so far published on *Prochloron*. Some inferences can be drawn about the evolution of *Prochloron* and its symbiosis with ascidians.

Cox, G., and D. M. Dwarte. Freeze-etch ultrastructure of a *Prochloron* species, the symbiont of *Didemnum molle*. New Phytol. 88:427–438; 1981.

A species of *Prochloron,* a genus of photosynthetic prokaryotes containing chlorophylls *a* and *b* and lacking phycobilins, has been investigated by the freeze-etch technique and thin sectioning. The thylakoids show a high level of stacking; particle distributions on fracture faces of the membranes resemble those of chloroplasts more than those of cyanophytes. Polyhedral bodies (carboxysomes?) are surrounded by a unit membrane which probably derives from the thylakoids. Vacuole-like areas are seen in freeze-etch replicas to contain a structured substance, probably a reserve material.

Critchley, C., and T. J. Andrews. Photosynthesis and plasmamembrane permeability properties of *Prochloron.* Arch. Microbiol. 138:247–250; 1984.

Photosynthetic carbon fixation of freshly isolated cells of *Prochloron,* the symbiont of *Lissoclinum patella,* proceeded at high rates (80–180 μmol $O_2 \cdot$ mg $Chl^{-1} \cdot h^{-1}$) in buffered seawater and showed a typical light response, saturating at about 300 $\mu E \cdot m^{-2} \cdot s^{-1}$. However, in NaCl solutions osmotically equivalent to seawater CO_2-dependent O_2 evolution ceased or was severely inhibited. Hypotonic or hypertonic conditions induce degrees of swelling or shrinkage, respectively, apparently causing similar increases in the plasmamembrane's permeability to ferricyanide. Initially high, but rapidly declining, rates of electron transport were observed when the cells were suspended in distilled water. This inhibition was not caused by rupture of the cells, indicating instead diffusive loss of some essential factor(s) which normally exchange easily and rapidly between the cells and/or the host environment. Such rapid exchange may be part of the mechanism of this symbiosis and, if not adequately understood, may frustrate attempts to culture *Prochloron* away from its host.

Fall, R., R. A. Lewin, and L. R. Fall. Intracellular coagulation inhibits the extraction of proteins from *Prochloron.* Phytochemistry 22:2365–2368; 1983.

Protein extraction from the prokaryotic alga *Prochloron* LP (isolated from the ascidian host *Lissoclinum patella*) was complicated by an irreversible loss of cell fragility in the isolated algae. Accompanying this phenomenon, which we term intracellular coagulation, was a redistribution of thylakoids around the cell periphery, a loss of photosynthetic O_2 production, and a drastic decrease in the extractability of cell proteins. Procedures are described for the successful preparation and transport of cell extracts yielding the enzymes glucose–6-phosphate dehydrogenase and 6-phosphogluconate dehydrogenase as well as other soluble proteins.

Fisher, C. R., Jr., and R. K. Trench. *In vitro* carbon fixation by *Prochloron* sp. isolated from *Diplosoma virens.* Biol. Bull. 159:636–648; 1980.

Prochloron sp. isolated from *Diplosoma virens* and incubated in the light in $NaH^{14}CO_3$ demonstrated a high photosynthetic capacity (up to 3.7 $\mu gC \cdot (\mu g$ Chlorophyll $a)^{-1} \cdot hr^{-1}$). *In vitro* these cyanobacteria release a maximum of 7% of the ^{14}C they fix in the light. Dark fixation was found to be maximally 3% of light fixation and release in the dark averaged 26% of the total ^{14}C fixed in the dark. These data imply that the organic carbon released by these cyanobacteria may not be quantitatively important to the host.

The labeled compound released by *Prochloron* in the light is glycolic acid. The major compounds produced by light and dark carbon fixation in *Prochloron* are identified, and similarities to other photosynthetic cyanobacteria are noted.

Florenzano, G., W. Balloni, and R. Materassi. Nomenclature of *Prochloron didemni* (Lewin, 1977) sp. nov., nom. rev., *Prochloron* (Lewin, 1976) gen. nov., nom. rev., *Prochloroceae* fam. nov., *Prochlorales* ord. nov., nom. rev. in the class *Photobacteria* (Gibbons and Murray, 1978). Int. J. Systematic Bacteriol. 36:351–353; 1986.

We propose that the photosynthetic procaryotes containing chlorophylls *a* and *b* in the species *Prochloron didemni* sp. nov., genus *Prochloron* gen. nov., be placed under the International Code of Nomenclature of Bacteria by including the genus *Prochloron* in the family *Prochloraceae* fam. nov., order *Prochlorales* ord. nov. in the class *Photobacteria* Gibbons and Murray 1978, listed on the approved lists of bacterial names.

Foss, P. A., R. A. Lewin and S. Liaaen-Jensen. Carotenoids of *Prochloron* sp. Phycologia 26:142–144; 1987.

Cells of *Prochloron* from *Lissoclinum patella* contained chlorophylls *a* and *b*, and carotenoids (0.34% of the acetone-extracted residue) comprising $\beta.\beta$ carotene, β,β carotene mono-epoxide, mutachrome, echinenone, isocryptoxanthin, (3R)-cryptoxanthin, (3R,3′R)-zeaxanthin, and an unidentified component (<1%).—*eds.*

Francis, S., E. S. Barghoorn, and L. Margulis. On the experimental silification of microorganisms. III. Implications of the preservation of the green prokaryotic alga *Prochloron* and other coccoids for interpretation of the microbial fossil record. Precambrian Res. 7:377–383; 1978.

Evidence for Precambrian fossil eukaryotic microorganisms has been based on: (1) the presence of internal 'spots' which have been variously interpreted to be remains of nuclei or pyrenoids of photosynthetic plastids or other organelles; (2) tetrahedral tetrad arrangement of cells; (3) trilete scars interpreted to be indicative of meiotic division: (4) large cell diameters; and (5) putative mitotic cell divisions. These features have been reported in fossils preserved in Precambrian cherts. We have studied modern microbial mats, thought to be analogues of Precambrian fossil communities, and found they may be silicified by laboratory procedures. In microbial mats from Baja California we have found many 'spot cells' that we could identify as remains of cyanophytes. We have silicified the newly discovered large prokaryotic coccoid green alga *Prochloron* and have found that it, like many cyanophytes previously silicified, preserves its structure and maintains its initial dimensions. In laboratory-silicified prokaryotic organisms we have found that all of the above criteria, supposedly characteristic of eukaryotes, can be observed. We conclude that there is no compelling morphological evidence for fossil eukaryotic microbes from Precambrian cherts.

Fredrick, J. F. The α-1,4-glucans of *Prochloron*, a prokaryotic green marine alga. Phytochemistry 19:2611–2613; 1980.

Prochloron, a symbiont found associated with *Lissoclinum patella* (a marine colonial ascidian), was lyophilized and its glucans extracted. The glucans were complexed with *s*-triazine reactive dyes and separated by electrophoresis on cellulose acetate membranes. A highly branched glucan similar to phytoglycogen, and a linear unbranched glucan resembling a short-chain amylose were both detected. This unusual polysaccharide mixture suggests a possible mode of starch biosynthesis in algae in general.

Fredrick, J. F. Glucosyltransferase isozymes forming storage glucan in *Prochloron*, a prokaryotic green alga. Phytochemistry 20:2353–2354; 1981.

Since the prokaryotic, green marine alga *Prochloron* has not, as yet, been cultured, lyophilized cells were used in a microadaptation of polyacrylamide gel electrophoresis (PAGE) in order to isolate the glucosyltransferase isozymes. The pattern obtained with these capillary gels was identical with those of cyanophytes. Besides two phosphorylase and synthase isozymes, three branching isozymes of the *b.e.* type were found to be present.

Giddings, T. H., Jr., N. W. Withers, and L. A. Staehelin. Supramolecular structure of stacked and unstacked regions of the photosynthetic membranes of *Prochloron* sp., a prokaryote. Proc. Natl. Acad. Sci. USA 77:352–356; 1980.

Freeze-fracture replicas of the photosynthetic prokaryote *Prochloron* sp., collected at Coconut Island, Hawaii, show that the thylakoids are differentiated into stacked and unstacked regions much like the thylakoids of green algae and higher plants. On the exoplasmic (E) fracture face, the particle density is greater in stacked regions (≈3100 particles/μm^2) than in unstacked regions (≈925 particles/μm^2). On the complementary protoplasmic (P) fracture face, the particle density is lower in stacked regions (≈2265 particles/μm^2 than in unstacked regions (≈4980 particles/μm^2). The histogram of the diameters of E face particles in unstacked regions shows a single broad peak centered at 80 Å. In stacked regions, four distinct peaks, at 75, 105, 130, and 160 Å, are observed. These size classes are virtually identical to those found on E faces of thylakoids of the green alga *Chlamydomonas* and of greening pea chloroplasts. In the latter systems, the different size classes of E face particles are believed to represent photosystem II units surrounded by variable amounts of chlorophyll *a/b* light-harvesting complex. We propose that the same interpretation applies to the thylakoids of *Prochloron*, which contain a similar chlorophyll *a/b* complex. Our results add to the evidence supporting the view of the chlorophyll *a/b* complex as both a light-harvesting complex and a membrane adhesion factor. The similarity of the architecture of the thylakoids of *Prochloron* to that of green algal and plant chloroplasts also provides additional evidence of an evolutionary relationship between *Prochloron* and the chloroplasts of green plants.

Griffiths, D. J., and L.-V. Thinh. Transfer of photosynthetically fixed carbon between the prokaryotic green alga *Prochloron* and its ascidian host. Aust. J. Mar. Freshw. Res. 34:431–440; 1983.

In the symbiotic association between the prokaryotic green alga *Prochloron* and three didemnid host species (*Diplosoma similis, Lissoclinum bistratum, Trididemnum cyclops*), between 6 and 51% of the total carbon fixed during exposure for 1 h to $H^{14}CO_3$ in the light (150 $\mu E\ m^{-2}\ s^{-1}$) becomes associated with the host tissue. Dark fixation of $^{14}CO_2$ in these ascidian species and in *Lissoclinum punctatum* never exceeds 6% of photosynthetic fixation at saturating light intensity. The corresponding values for dark fixation of $^{14}CO_2$ in isolated *Prochloron* cells fall within the same range. There is very little excretion of photosynthate from whole colonies of the above ascidian species nor from *Didemnum molle, Lissoclinum voeltzkowi* and *Trididemnum miniatum* (usually less than 1% of total photosynthate at saturation light intensity), suggesting an efficient transfer mechanism from *Prochloron* to host. Evidence from pulse–chase experiments suggests that transfer probably involves the early products of photosynthesis. The extent of transfer of photosynthate between *Prochloron* and *T. cyclops* varies with the rate of photosynthetic $^{14}CO_2$ fixation into the whole colony but there is some transfer even at low light intensities, which strongly limit photosynthesis.

Griffiths, D. J., and L.-V. Thinh. Photosynthesis by *in situ* and isolated *Prochloron* (Prochlorophyta) associated with didemnid ascidians. Symbiosis 3:109–122; 1987.

Prochloron cells isolated from their symbiotic association with the didemnid ascidian *Trididemnum cyclops* have a light-saturated net photosynthetic rate of 148 μmol O_2 mg chl*a*$^{-1}$ h^{-1}. Light saturated *in situ* photosynthesis occurs at approximately 32% of this rate. Comparable results were obtained with isolated and *in situ Prochloron* associated with small colonies of *Lissoclinum patella* and with discs cut from large colonies. In the latter, prolonged washing (42 hr) reduced *in situ* photosynthesis to 7% of the photosynthetic capacity of corresponding released algal cells. *In situ* photosynthesis, for both associations, incorporates a greater proportion of photosynthetically-fixed carbon into the algal protein fraction (c. 20%) than occurs from photosynthesis of isolated *Prochloron* (<5%). For both associations, approximately 20% of photosynthetically-fixed carbon is transferred to the host, but only in *L. patella* was there significant incorporation into the TCA-insoluble fraction, much of it associated with the test. It is concluded that restriction of *in situ* photosynthesis by the host helps to match algal proliferation to the growth rate of the host.

Griffiths, D. J., L.-V. Thinh, and H. Winsor. Crystals and paracrystalline inclusions of *Prochloron* (Prochlorophyta) symbiotic with the ascidian *Trididemnum cyclops* (Didemnidae). Botanica Marina 27:117–128; 1984.

Crystals and paracrystalline inclusions are described from cells of the green prokaryotic alga *Prochloron* symbiotic with the tropical ascidian *Trididemnum cyclops*. These include polyhedral bodies (which bear a close resemblance to carboxysomes described from other photosynthetic prokaryotes), large crystals (resembling protein crystals of higher plant chloroplasts), spherical bodies (thought to be a different form of the large crystals), ordered lamellae and cylindrical bundles. It is suggested that all these inclusions may be involved with temporary storage of components of the photosynthetic system. They highlight the similarity of *Prochloron* to the other photosynthetic prokaryotes but separate it from the eukaryotic algal groups.

Herdman, M. Deoxyribonucleic acid base composition and genome size of *Prochloron*. Arch. Microbiol. 129:314–316; 1981.

The DNA base composition of the photosynthetic prokaryote *Prochloron* was determined (on samples collected from the natural environment) to be 40.8 mol % GC. The sharp differential melting curve indicated the absence of significant quantities of contaminating DNA from other organisms. The genome size, estimated from the renaturation kinetics of thermally denatured DNA, was 3.59×10^9 daltons mol. wt, similar to that of many other prokaryotes. The fact that *Prochloron* has not yet been cultured in the laboratory cannot, therefore, be attributed to a reduced genetic information content.

Hiller, R. G., and A. W. D. Larkum. The chlorophyll-protein complexes of *Prochloron* sp. (Prochlorophyta). Biochimica et Biophysica Acta 806:107–115; 1985.

Chlorophyll-protein complexes have been isolated from *Prochloron* sp. by SDS-polyacrylamide gel electrophoresis and SDS-sucrose-gradient centrifugation. Complexes associated with Photosystem I have significant amounts of chlorophyll *b* and a principle polypeptide of 70 kDa. The largest Photosystem I complex had an M_r of more than 300 000 kDa, a chlorophyll *a/b* ratio of 3.8 and a chlorophyll/P–700 ratio of approx. 100. Complexes enriched in chlorophyll *b* showed reduced electrophoretic mobility compared to spinach LHCP3, a higher Chl *a/b* ratio (approx. 2.4) and had a principle polypeptide of 34 kDa. Neither the 34 kDa or any other polypeptide showed cross-reactivity with antibodies to spinach light-harvesting chlorophyll *a/b* protein in a Western blot test.

Johns, R. B., P. D. Nichols, F. T. Gillan, G. J. Perry, and J. K. Volkman. Lipid composition of a symbiotic prochlorophyte in relation to its host. Comp. Biochem. Physiol. 69B:843–849; 1981.

1. Two samples of the ascidian *D. molle* (grey and brown forms), *Prochloron sp.* cells, and a plankton sample have been analyzed for carotenoids, fatty acids and sterols in a study directed towards determining energy flow in the symbiotic prochlorophyte-ascidian system.
2. The near identity of carotenoid compositional patterns between the *Prochloron sp.* and the sample of *D. molle* taken together with the relative absolute concentrations suggest a 2–3% biomass contribution of prochlorophyte to the ascidian samples.
3. The data from sterol analyses suggest that the prochlorophyte does not biosynthesize its own sterols.
4. A fatty acid analysis of *D. molle* samples show 16:0, 16:1ω7, 20:4ω6 and 20:5ω3 to be the major components, and there is reason to believe that the polyunsaturated fatty acids may be derived, at least in part, from a plankton food source.
5. Sterol analyses of the ascidian samples suggest that this animal can reduce Δ^5-sterols, which may also be derived through a plankton diet, to 5α-stanols, and can convert Δ^5-sterols to the Δ^7 isomers.

Kenrick, J. R., E. M. Deane, and D. G. Bishop. A comparative study of the fatty acid composition of *Prochloron* lipids. Phycologia 23:73–76; 1984.

The major lipids of cells of the symbiotic procaryote *Prochloron*, isolated from five different hosts, are monogalactosyldiacylglycerol, digalactosyldiacylglycerol, phosphatidylglycerol and sulphoquinovosyldiacylglycerol. The major fatty acids of these lipids are saturated or mono-unsaturated acids containing 14 or 16 carbon atoms. Marked variations occur in the fatty acid composition of the galactolipids isolated from *Prochloron* cells from the same host at different locations or at different times. These variations appear to preclude the use of fatty acids as a taxonomic criterion for differentiating species of *Prochloron*.

Kremer, B. P. Prokaryotische Grünalgen—Aspekte zur Biologie der Prochlorophyta. Biol. Rdsch. 22:277–288; 1984.

Bright green, unicellular algae have been reported to be symbiotically associated with ascidians from tropical Pacific shores, particularly with the members of the family Didemnidae. The cells are definitely prokaryotic. However, they lack phycobiliproteins, but contain fair amounts of chlorophyll *b* along with chlorophyll *a* instead. An assignment of such cells to the hitherto recognized algal divisions therefore presented a major taxonomic problem. To accommodate them taxonomically, a new division, Prochlorophyta, was proposed. The respective species were transferred to *Prochloron* as the type genus. Some peculiar features of *Prochloron* with regard to cell morphology, biochemistry, physiology, and phylogeny are reviewed and discussed from a comparative viewpoint.

Kremer, *B. P., R. Pardy, and R. A. Lewin. Carbon fixation and photosynthates of Prochloron,* a green alga symbiotic with an ascidian, *Lissoclinum patella.* Phycologia 21:258–263; 1982.

It is estimated that *Prochloron* cells, freshly isolated from colonies of their ascidian host, *Lissoclinum patella,* can fix CO_2 photosynthetically at rates exceeding 120 m*M* g^{-1} chlorophyll h^{-1}. In longer fixation periods (up to 1 h) proportionately more insoluble material was formed. Of the products soluble in aqueous alcohol, more than 50% consisted of amino acids and about 20% of acids in the TCA cycle, additionally including glycolate. Free monosaccharides or disaccharides are lacking. The lipid products included non-polar lipids (24%), diglycerides (37% monogalactosyl) and phosphatidyl glycerol (11%). Oligoglucans composed more than 95% of the residual insoluble fraction. The results are discussed from a chemotaxonomic point of view.

Lewin, R. A. A marine *Synechocystis* (Cyanophyta, Chroococcales) epizoic on ascidians. Phycologia 14:153–160; 1975.

Synechocystis didemni spec. nov. is described. It has only been found growing epizoically on calcified colonial ascidians (*Didemnum* spp.) in subtropical coastal waters. The apparent absence of phycobilin pigments and the presence of two chlorophylls (separable by chromatography) are unusual features for a photosynthetic prokaryont. So far, all attempts to culture this alga have been unsuccessful.

Lewin, R. A. Prochlorophyta as a proposed new division of algae. Nature 261:697–698; 1976. [Full text]

Unicellular algae associated with ascidians from tropical Pacific shores have been reported by various biologists.[1–5] They are bright green, generally spherical and about 10 20μm in diameter, and they seem to have no clearly delimited nucleus or plastids. Such cells (identified as *Synechocystis didemni*), found associated with surfaces of *Didemnum* colonies on the Pacific coast of Mexico, have been shown by electron microscopy to be prokaryotic,[6, 7] which suggests that they are cyanophytes, that is, blue-green algae. Although all known blue-green algae (other than a few apochlorotic types) contain phycoerythrin, phycocyanin, or both, however, these ascidian symbionts are apple green and contain no detectable bilin pigments. Furthermore, like the eukaryotic algae in the divisions Chlorophyta and Euglenophyta, they contain two chlorophyll components, separable by chromatography and provisionally identifiable as chlorophylls *a* and *b*[8], whereas no cyanophytes are known to contain chlorophyll *b*. The assignment of *S. didemni* to any of the established algal divisions, therefore, presents a major taxonomic problem.

Algal cells, found living inside the colonies of other didemnid ascidians at Enewetak Atoll, Marshall Islands, and on Coconut Island, Hawaii, have been recently studied by the same techniques and have proved to be very much like those of *S. didemni*. My colleagues and I (unpublished results) have confirmed in particular that those from *Diplosoma virens* (from Hawaii) are prokaryotic, contain chlorophylls *a* and *b*, contain no detectable bilin pigments (even after incubation for one or more days in nitrogenous media) and show no evidence of phycobilisomes on the thylakoids. In the light, in aerobic conditions, they can fix CO_2 and evolve oxygen. They must therefore be considered as algae (rather than as photosynthetic bacteria, which are normally anaerobic and, when illuminated, fix CO_2 without concomitant oxygen evolution). They cannot, however, be logically assigned to the Cyanophyta, the Chlorophyta, the Euglenophyta, or any other existing division of the algae.

In view of the fact that their peculiar combination of characteristics is common to algal symbionts from several different ascidians found in widely separated locations, it now seems appropriate to put such organisms in a new algal division, which I propose to call the Prochlorophyta. Since it will consequently be necessary to remove the type species, *S. didemni*, from the Cyanophyta, it can no longer be retained in the blue-green algal genus *Synechocystis*, in spite of morphological similarities. The new division is formally described in Latin (requisite for all new plant taxa) as follows: *Prochlorophyta: algae procarioticae similes Cyanophytis, cuiis differentes pro formatio chlorophyllarum binarum (a, b-que); pigmentia rutila aut caerulea absunt.*

(Formal descriptions of the new class, order, family, genus and type species will be presented elsewhere.)

RALPH A. LEWIN

Scripps Institution of Oceanography,
University of California,
La Jolla, California 92093

Received April 2; accepted April 26, 1976.

[1]Smith, H. G., *Ann. Mag. nat. Hist.*, 15, 615 (1935).
[2]Eldredge, L. G., *Micronestra.* 2, 161 (1967).
[3]Tokioka, T., *U.S. natn. Mus. Bull.*, 251, 247 pp (1967).
[4]Newcomb, E. H., and Pugh, T. D., *Nature*, 253, 533 (1975).
[5]Lewin, R. A., and Cheng, L., *Phycologia*, 14, 149 (1975).
[6]Lewin, R. A., *Phycologia*, 14, 153 (1975).
[7]Schulz-Baldes, M., and Lewin, R. A., *Phycologia*, 15 (1976).
[8]Lewin, R. A., and Withers, N. W., *Nature*, 256, 735 (1975).

Lewin, R. A. *Prochloron*, type genus of the Prochlorophyta. Phycologia 16:217; 1977. [Full text]

Certain unicellular, green-coloured algae associated with didemnid ascidians of Pacific shores have been found to be prokaryotic, like blue-green algae, but to have photosynthetic pigments like those of green algae. Since they therefore cannot reasonably be assigned to the Cyanophyta, the Chlorophyta, or any other accepted division of the algae, a new division, the Prochlorophyta, has been proposed for such plants (Lewin, 1976). The type species, hitherto regarded as a blue-green alga and named accordingly as *Synechocystis didemni*, must now be given a new generic name, to constitute the basis for the establishment of a new family, order and class. At this stage, descriptions must necessarily be brief: one cannot predict what other kinds of prochlorophytes—perhaps now masquerading as blue-green algae—may be discovered in the future. Until we know more about possible specific differences among the algal associates of the various ascidians which harbour them, and pending the possible discovery of other prokaryotic algae which possess chlorophylls *a* and *b* but lack bilin pigments, we cannot reasonably specify distinguishing features of the new taxa. The following brief descriptions are proposed.

Division. Prochlorophyta (Lewin, 1976). Prokaryotic algae resembling cyanophytes, from which they differ in that they form chlorophylls *a* and *b* and lack accessory red or blue bilin pigments.

Class. Prochlorophyceae. With the characters of the division.

Order. Prochlorales. Unicellular or filamentous, branched or unbranched.

Family. Prochloraceae. Unicellular, encapsulated or not.

Genus. *Prochloron*. Unicellular, spherical, without evident mucilaginous sheath. Only form of reproduction so far observed: binary division, by equatorial constriction. So far, only found as extracellular symbionts of ascidians on marine shores.

Species. *Prochloron didemni*
(= *Synechocystis didemni* R. A. Lewin, 1975)
This is the type species for the genus, family, order, class and division.

Formal descriptions in Latin are appended, in accordance with the International Code of Botanical Nomenclature.

Prochlorophyceae:	sicut pro divisione.
Prochlorales:	unicellulares aut filamentosae, ramosae aut non ramosae.
Prochloraceae:	unicellulares, capsulatae aut sine capsula mucilaginosa visibili.
Prochloron:	unicellulares, sphaericae, sine capsula mucilaginosa visibili. Usque adhuc unus solum reproductionis modus cognitus, id est, divisio in duas partes per constrictionem aequatorialem. Usque adhuc inventae solum ut cellulae in coloniis ascidiarum conviventes in oris maritimis.

References

LEWIN, R. A. (1975.) A marine *Synechocystis* (Cyanophyta, Chroococcales) epizoic on ascidians. *Phycologia*, 14, 153–160

LEWIN, R. A. (1976.) Prochlorophyta as a proposed new division of algae. *Nature, Lond.*, 261, 697–698.

Ralph A. Lewin, Scripps Institution of Oceanography, University of California, La Jolla, CA. 92093, U.S.A.

Lewin, R. A. Distribution of symbiotic didemnids associated with prochlorophytes. Proc. of the Intl. Symposium on Marine Biogeography and Evolution in the Southern Hemisphere, Auckland, NZ, 17–20 July 1978, 2:365–369 (NZ DSIR Information Series 137).

Didemnids with intracolonial prochlorophytes occur on many tropical shores where temperatures remain above 21°C throughout the year. Prochlorophytes also occur sporadically on the surfaces of didemnid colonies growing in somewhat cooler situations, but they have not been recorded outside the tropics.

Lewin, R. A. The Prochlorophytes. *In:* The Prokaryotes: A handbook on habitats, isolation, and identification of Bacteria (M. P. Starr, H. Stolp, H. G. Trüper, A. Balows, and H. G. Schlegel, eds.), 1 Springer-Verlag; 1981:257–266.

Although *Prochloron* has not yet been successfully grown in culture, a body of information has accrued on the basis of studies of natural collections preserved in various ways. This review summarizes information published up to 1979.—*eds.*

Lewin, R. A. *Prochloron* and the theory of symbiogenesis. Annals of the New York Academy of Sciences 361:325–329; 1981.

The discovery of *Prochloron* allows us to test some of the assumptions underlying Mereshkovsky's theory of symbiogenesis, whereby green-plant chloroplasts may have arisen from intracellular prokaryotic algae.—*eds.*

Lewin, R. A. The problems of *Prochloron*. Ann. Microbiol. (Inst. Pasteur) 134B:37–41; 1983.

Prokaryotic green algae (prochlorophytes), which contain chlorophylls *a* and *b* but no bilin pigments, may be phylogenetically related to ancestral chloroplasts if symbiogenesis occurred. They may be otherwise related to eukaryotic chlorophytes. They could have evolved from cyanophytes by loss of phycobilin and gain of chlorophyll *b* synthesis. These possibilities are briefly discussed. Relevant evidence from biochemical studies in many collaborative laboratories is not becoming available for the resolution of such questions.

Lewin, R. A. *Prochloron*—A status report. Phycologia 23:203–208; 1984.

Prochloron is a genus of prokaryotic algae with photosynthetic pigments like those of chlorophytes. Prochlorophytes are almost invariably found associated as symbionts with marine protochordates (didemnid ascidians), and so far none has been successfully grown in sustained culture away from its host. Based on materials collected from nature, information of various sorts (biochemical, physiological, cytological and fine-structural) has been obtained, indicating many resemblances (and probably close phylogenetic affinities) between prochlorophytes and cyanophytes. Nevertheless they are distinguished by certain unique combinations of characters. Some of the data support the symbiogenesis theory for the origin of green-plant chloroplasts. Other possibilities are briefly discussed.

Lewin, R. A. The phylogeny of *Prochloron*. Giornale Botanico Italiano 120:1–14; 1986.

Prochloron, a unicellular alga that combines some features of cyanophytes with others of chlorophytes, is a phylogenetic enigma. Mounting evidence from electron microscopy, comparative biochemistry and molecular biology now suggests that prochlorophytes probably arose from blue-green algal ancestors, perhaps less than 10^8 years ago.

Lewin, R. A., and L. Cheng. Associations of microscopic algae with didemnid ascidians. Phycologia 14:149–152; 1975.

Microscopic epizoic algae were found, forming coloured patches on the surfaces of calcified colonial ascidians (*Didemnum* spp.) along shores of the Gulf of California.

The commonest of these is an apparently undescribed species of *Synechocystis*. Other algae more rarely associated with these ascidians include pink unicellular and filamentous cyanophytes and diatoms.

Lewin, R. A., L. Cheng, and R. S. Alberte. *Prochloron*–ascidian symbioses: photosynthetic potential and productivity. Micronesica 19:165–170; 1983.

We determined the chlorophyll content of didemnid ascidians with symbiotic algae (*Prochloron* sp.) from tropical marine waters around Palau, Western Caroline Islands. Several species contain as much chlorophyll per unit dry weight as many herbaceous crop plants and more than other symbiotic associations such as lichens, green hydra, etc. Their chlorophyll *a*/*b* ratios (3–9) were generally much higher than those of angiosperms (2–4). Where they abound, *Prochloron*–ascidian symbioses could make a major contribution to the productivity, especially in localized areas of tropical marine waters characterized by low nutrient levels and high irradiance.

Lewin, R. A., L. Cheng, and R. S. Alberte. Ecology of *Prochloron*, a symbiotic alga in ascidians of coral reef areas. Proc. Fifth Int. Coral Reef Cong., Tahiti, 1985; 5:95–101; 1985.

Prochloron is an unusual prokaryotic, unicellular alga combining features of both chlorophytes and cyanophytes. It is normally confined to symbiotic associations with colonial ascidians in the family Didemnidae, which grow on illuminated surfaces in low-intertidal or subtidal zones on tropical or subtropical seashores. The algal cells occur in the cloacal system of the host or embedded in its "test." They are usually outside the host cells, but in one species (*Lissoclinum punctatum*) many are intracellular. Certain host species have evolved special adaptations for this symbiosis, including a device whereby algal cells are attached to the free-swimming larvae. *Prochloron* cells, even when isolated from their hosts, can photosynthesize at rates comparable to those of green eukaryotic algae. However, they have not yet been cultured successfully in the laboratory. The host presumably provides essential growth factors or otherwise adapts the microenvironment to favor its algal symbionts. After photosynthesis, *Prochloron* liberates soluble organic products, including glycolate: uptake of such solutes by host tissues has been demonstrated. In one host species (*Liss. patella*) the symbionts have also been shown to fix molecular nitrogen. We have tabulated some comparative physiological data for the symbiotic algae from different host didemnids and discuss their implications for the ecology of both symbionts.

Lewin, R. A., L. Cheng, and F. Lafargue. Prochlorophytes in the Caribbean. Bull. of Mar. Sci. 30:744–745; 1980.

Prochloron is recorded in the Caribbean: in the Virgin Islands, Puerto Rico, and Grand Cayman.—*eds.*

Lewin R. A., L. Cheng, and J. Matta. Diurnal rhythm in the cell-division frequency of *Prochloron* (Prochlorophyta) in nature. Phycologia 23:505–507; 1984.

Cell division stages of *Prochloron* in three host didemnids may reach up to 15 percent during the morning and early afternoon and decline to 4 percent or less at night.—*eds.*

Lewin, R. A., and N. W. Withers. Extraordinary pigment composition of a prokaryotic alga. Nature 256:735–737; 1975.

A prokaryote given the name *Synechocystis* (now = *Prochloron*) *didemni* is reported to contain chlorophylls *a* and *b*, β-carotene, and at least three xanthophylls, but no demonstrable water-soluble phycobilin pigment.—*eds.*

MacKay, R. M., D. Salgado, L. Bonen, E. Stackebrandt, and W. F. Doolittle. The 5S ribosomal RNAs of *Paracoccus denitrificans* and *Prochloron*. Nucleic Acids Res. 10:2963–2970; 1982.

The nucleotide sequences of the 5S rRNAs of *Paeracoccus denitrificans* and *Prochloron* sp. are pGUCUGGUGGCCAAAGCACGAGCAAAACACCCGAUCCA UCCCGAACUCGGCCGUUAAGUGCCGUAGCGCCAAUGGUACUGCGU CAAAAGACGUGGGAGAGUAGGUCACCGCCAGACC$_{OH}$ and UUCCUG GUGUCUCUAGCGCUUUGGAACCACUUCGAUUCCAUCCCGAACUCG AUUGUGAAACUUUGCUGCGGCUAAGAUACUUGCUGGGUUGCUGGC UGGGAAAAUAGCUCGAUGCCAGGAUU$_{OH}$, respectively. Specific phylogenetic relationships of *P. denitrificans* with purple non-sulphur bacteria, and of *Prochloron* with cyanobacteria are demonstrated, and unique features of potential secondary structure are described.

(Position numbers printed beneath the sequences: 10, 20, 30, 40, 50, 60, 70, 80, 90, 100, 110; 10, 20, 30, 40, 50, 60, 70, 80, 90, 100, 110, 120.)

McCourt, R. M., A. F. Michaels, and R. W. Hoshaw. Seasonality of symbiotic *Prochloron* (Prochlorophyta) and its Didemnid host in the northern Gulf of California. Phycologia 23:95–101; 1984.

Prochloron didemni (Lewin) white Lewin, a prokaryotic unicell grows in films on the external surfaces of part of the population of the colonial tunicate *Didemnum* sp. in intertidal areas of the northern Gulf of California. Seasonal abundance of *P. didemni* and its host were monitored for 13 months at Puerto Peñasco, Sonora, Mexico. *Prochloron didemni* was most abundant in summer. In months when water temperatures were above 21°C, the number of *Didemnum* colonies with attached *Prochloron* correlated significantly with sea surface temperature. The number of colonies with *P. didemni* was at a minimum in winter and was not significantly correlated with temperatures less than 21°C. Abundance of *Didemnum* colonies varied independently of *P. didemni* abundance and showed no apparent seasonal cycle. Individual colonies frequently were overgrown by other sessile organisms or buried by sand. *Prochloron didemni* was more commonly found on large host colonies than small, although no more than 43% of *Didemnum* colonies had visible *P. didemni* on them during the period of its peak abundance (July–August). We conclude that temperature or a correlated factor such as amount of sunlight, and not available surface space on host colonies, controls *P. didemni* abundance in this region.

Moriarty, D. J. W. Muramic acid in the cells walls of *Prochloron*. Arch. Microbiol. 120:191–193; 1979.

Muramic acid has been detected in *Prochloron* with the aid of two different techniques. It was assayed by cleaving D-lactate from muramic acid and then reducing NAD with D-lactate dehydrogenase and measuring the NADH with bacterial luciferase. Gas-liquid chromatography of trimethylsilyl derivatives of cell extracts confirmed that muramic acid was present in about the quantity given by the D-lactate assay. The amount of muramic acid present was 1.7 ± 0.2 μg/mg dry weight or 1.3 fg/μm^2 of cell surface. This suggests that the thickness of the peptidoglycan layer in *Prochloron* is similar to that in blue-green algae.

Müller, W. E. G., A. Maidhof, R. K. Zahn, J. Conrad, T. Rose, P. Stefanovich, I. Müller, U. Friese, and G. Uhlenbruck. Biochemical basis for the symbiotic relationship *Didemnum-Prochloron* (Prochlorophyta). Biol. Cell 51:381–388; 1984.

Prochloron cells (prochlorophytes), living in symbiosis with colonial ascidians, have been isolated from *Didemnum molle* (collected around Giravaru, Maldives) and from *Didemnum fulgens* (gathered around Rovinj, Yugoslavia). The identification of the *Prochloron* samples is based on morphological, immunological and biochemical criteria. The growth state of *D. molle* was found to be correlated with the amount of algal symbionts which are associated with the invertebrates. Methanol extracts from *D. molle* displayed strong cytostatic activity if tested in cultures with L5178y mouse lymphoma cells; division of *Prochloron* cells was not influenced by this cytostatic activity. Colonies of *D. molle* which are rich in these algae produce large amounts of a lectin (1,140 hemagglutination units/mg). Colonies of *D. molle* lacking *Prochloron* did not contain any lectin activity. The lectin was determined to be a component required for a successful *in vitro* cultivation of *Prochloron*. From the presented data we conclude that the symbiotic relationship between the ascidians and *Prochloron* is based on the presence of cytostatic compounds (produced by the animal) and on the transfer of nutriments from the algae to the host. Furthermore, this symbiotic relationship seems to be stabilized and perhaps even maintained by a lectin.

Murata, N., and N. Sato. Analysis of lipids in *Prochloron* sp.: occurrence of monoglucosyl diacylglycerol. Plant & Cell Physiol. 24:133–138; 1983.

Prochloron contained monogalactosyl diacylglycerol, digalactosyl diacylglycerol, sulfoquinovosyl diacylglycerol, phosphatidylglycerol and, as a minor component, monoglucosyl diacylglycerol, but no phosphatidylcholine. With respect to the lipid and fatty acid compositions, this alga is similar to the blue-green algae rather than the chloroplasts of eukaryotic plants.

Newcomb, E. H., and T. D. Pugh. Blue-green algae associated with ascidians of the Great Barrier Reef. Nature 253:533–534; 1975.

Cloacal pockets in blue colonies of *Diplosoma virens* [= *Dip. similis*], green colonies of *Lissoclinum molle* [= *L. bistratum*], and brown colonies of *Didemnum ternatanum* [= *Did. molle*] contained spherical algal cells (cf. *Anacystis aeruginosa*), respectively, 5–15, 20, and 7–15 μm in diameter. The algae were bright green, their absorption spectra indicating the presence of chlorophyll *a* and little or no phycobilin. No nitrogen fixation could be demonstrated. (Specific names of didemnids have been revised by Kott [1982; see Table 2].)—*eds.*

Olson, R. R. Light-enhanced growth of the ascidian *Didemnum molle*/*Prochloron* sp. symbiosis. Mar. Biol. 93:437–442; 1986.

To obtain a direct measurement of the importance of *Prochloron* sp. to the ascidian *Didemnum molle,* ascidian colonies from Lizard Island Lagoon, Great Barrier Reef, were grown for 9 d at 0, 10, 40, and 100% sunlight *in situ* using unidirectional flow chambers. Growth (wet weight) was enhanced up to 40% of full sunlight, at which point growth appears to have been light-saturated. Colonies in 10 and 40% sunlight responded by (1) climbing up the sides of the growth chambers, and (2) flattening out to a more encrusting morphology; also (3) the chlorophyll content of three colonies in zero sunlight decreased by >80%, yet the ascidians remained healthy and did not lose weight. These data show that although the symbiosis may not be obligatory for *D. molle,* the ascidian's growth is enhanced by *Prochloron* sp., and the morphology of the ascidian colony is affected by its photobiology.

Olson, R. R., and J. W. Porter. *In-situ* measurement of photosynthesis and respiration in the ascidian-*Prochloron* symbiosis *Didemnum molle.* Coral Reef Symposium, Tahiti, April 1985.

Photosynthesis and respiration of the colonial ascidian–algal (*Prochloron*) symbiosis *Didemnum molle* were measured in situ at 5 m depth over 24 h using a submersible respirometer. Several colonies were incubated within each of three chambers. By the end of 24 hours all colonies had attached to the base of the chambers and were fully expanded, indicating vigor. Average night time respiration was 3.32 (± 0.32 s.d.) mg 0_2 g dry wt^{-1} hr^{-1}. The ratio of photosynthesis to respiration (P/R) over 24 hours was 0.62. Colonies attained 95 percent light saturation ($I_{0.95}$) at 687 E m^{-2} s^{-1}. The percent contribution of *Prochloron* to ascidian respiration (CPAR) is calculated to be between 12 and 31 percent assuming translocation rates of 20 and 50 percent, respectively. While *Prochloron* can provide a significant portion of the ascidian's carbon requirement, no accurate data exist on the translocation efficiency of this symbiosis. Considering the location of the algae within the colony, in the excurrent chamber, the algae may provide only a small amount of nutrition.

Omata, T., M. Okada, and N. Murata. Separation and partial characterization of membranes from *Prochloron* sp. Plant Cell Physiol. 26:579–584; 1985.

Two differently colored membrane preparations were separated from the prochlorophyte, *Prochloron* sp., by mechanical disintegration of the cells followed by sucrose density gradient centrifugation. An orange-colored preparation, containing zeaxanthin as the major constituent pigment, seemed to comprise the cytoplasmic membrane. The other green-colored membrane preparation, containing β-carotene and chlorophyll *a* and *b* as major pigment constituents, was identified as the thylakoid membrane. The two types of membranes were compared as to their absorption spectra and buoyant densities.

Paerl, H. W. N_2 fixation (nitrogenase activity) attributable to a specific *Prochloron* (Prochlorophyta)–ascidian association in Palau, Micronesia. Mar. Biol. 81:251–254; 1984.

Light-mediated nitrogenase activity (NA) consistently occurred in one of four species of tropical marine ascidians (sea squirts) hosting the symbiotic alga *Prochloron* (prochlorophyta) among reef habitats located in Palau, Micronesia. NA was limited to intact colonies of *Lissoclinum patella* exclusively colonized by cells of *Prochloron.* Neither isolated viable cells of *Prochloron* nor ascidians free of *Prochloron* revealed NA, indicating a strong dependence on the intact symbiosis for creating conditions conductive to N_2 fixation. The confinement of NA to *L. patella* may be related to both the oligotrophic habitat and the large colony size of this protochordate species. This is the first report of NA attributable to symbiotic alga residing in either an aquatic or a terrestrial animal.

Paerl, H. W., R. A. Lewin, and L. Cheng. Variations in chlorophyll and carotenoid pigmentation among *Prochloron* (Prochlorophyta) symbionts in diverse marine ascidians. Botanica Marina 27:257–264; 1984.

High-performance liquid chromatography (HPLC) was used to separate and quantify chlorophylls *a* and *b* as well as major carotenoid pigments present in freeze-dried preparations of diverse *Prochloron*–didemnid associations and in *Prochloron* cells separated from host colonies. Both chlorophyll *a* and *b* were consistently observed in each association. Chlorophyll *a*:*b* ratios ranged from 4.14. to 19.71. Ratio differences reflected species differences among didemnid hosts, which proved consistent over time and space. Generally good agreement was found between ratios determined in isolated cell preparations and in symbiotic colonies (*in hospite*). These values are 1.5 to 5-fold higher than ratios determined in a variety of eukaryotic green plants. The carotenoids in *Prochloron* are quantitatively and qualitatively similar to those found in various freshwater and marine cyanophytes from high-light environments. However, *Prochloron* differs from most cyanophytes in the absence of myxoxanthophyll and related glycosidic carotenoids. The consistent presence of chlorophyll *b* and individuality in carotenoid pigmentation render *Prochloron* biochemically distinct from cyanophytes, despite the fact that cells of both prokaryotic groups are often found in light-saturated environments.

Pardy, R. L. Oxygen consumption and production by tropical ascidians symbiotic with *Prochloron*. Comp. Biochem. Physiol. 79A:345–348; 1984.

1. Oxygen consumption and production rates were measured in two species of colonial ascidians that contained the algal symbiont, *Prochloron*.
2. Despite differences in size and habitats, the colonies showed similar rates of oxygen consumption and production.
3. Oxygen production by the colonies was light dependent.
4. Based on the data presented, the symbiosis is similar to other algal–invertebrate symbioses in producing more oxygen than is consumed when illuminated.

Pardy, R. L., and R. A. Lewin. Colonial ascidians with prochlorophyte symbionts: evidence for translocation of metabolites from alga to host. Bull. Mar. Sci. 31:817–823; 1981.

We measured rates of incorporation of $^{14}CO_2$ by two species of tropical colonial ascidians (*Diplosoma* sp. and *Lissoclinum patella*) symbiotically associated with the unicellular alga *Prochloron*. Colonies incubated in light for 1 h incorporated 4–5 times as much acid-stable ^{14}C in the animal tissue as dark controls. We concluded that the animals took up photosynthate from their algal symbionts. *Prochloron* illuminated with $^{14}CO_2$ *in vitro* liberated small amounts of acid-stable ^{14}C compounds into the medium. Most of the ^{14}C in the host was found in molecules of low molecular weight; label was also detected in lipid, nucleic acid and protein fractions.

Pardy, R. L., R. A. Lewin, and K. Lee. The *Prochloron* symbiosis. *In:* Algal symbiosis: A continuum of interaction strategies (L. J. Godd, ed.), Cambridge University Press; 1983:91–96.

Prochloron cells can be easily squeezed out of cloacal cavities in colonies of *Lissoclinum patella*. Electron microscopy shows that the cell wall is like that of a cyanophyte. The thylakoids, which tend to be peripheral, are associated with polyhedral bodies.—*eds.*

Parry, D. L. Nitrogen assimilation in the symbiotic marine algae *Prochloron* spp. Mar. Biol. 87:219–222; 1985.

Prochloron spp. Lewin, 1977, from three species of ascidians [*Lissoclinum patella* (Gottschaldt, 1898), *Tridideinnum cyclops* (Michaelsen, 1921) and *Diplosoma virens* (Hartmeyer, 1909)] showed light-dependent assimilation of ^{15}N-labeled ammonium sulphate. The ^{15}N assimilated into glutamine was measured using the non-invasive technique of nitrogen-15 nuclear magnetic resonance (^{15}N NMR).

Parry, D. L. *Prochloron* on the sponge *Aplysilla* sp. Bull. Mar. Sci. 38:388–390; 1986.

Green patches on surfaces of an orange encrusting sponge, *Aplysilla* sp., at Heron Island, Great Barrier Reef, are attributable to epizoic *Prochloron* cells. They are 10–15 μm in diameter. Their chlorophyll *a*:*b* ratio, 2.49, is similar to that of *Prochloron* from didemnids.—*eds.*

Patterson, G. M., and N. W. Withers. Laboratory cultivation of *Prochloron,* a tryptophan auxotroph. Science 217:1934–1935; 1982.

Laboratory cultures have been established of the didemnid symbiont *Prochloron,* a unique prokaryotic alga that synthesizes chlorophylls *a* and *b* but no phycobilin pigments. Cell division in *Prochloron* cultures occurs under acidic conditions (pH 5.5) in the presence of tryptophan. The alga is a naturally occurring tryptophan auxotroph that survives in nature by close association with the host, *Diplosoma similis.* The metabolic dysfunction that renders *Prochloron* auxotrophic may involve only the initial step of the tryptophan biosynthetic pathway.

Perry, G. J., F. T. Gillan, and R. B. Johns. Lipid composition of a prochlorophyte. J. Phycol. 14:369–371; 1978.

A lipid analysis is reported of a symbiont *Prochloron* sp. associated with the ascidian *Lissoclinum patella* collected from Rodda Reef, North Queensland, Australia. Phosphatidyl glycerol, monogalactosyl diglyceride, digalactosyl diglyceride and sulfoquinovosyl diglyceride were identified. These lipids together with their fatty acid profiles, the limited range of hydrocarbons (nC_{15}–nC_{17}), and sterol analysis (cholesterol 27%) are consistent with a firm relationship for the new genus *Prochloron* with the Cyanophyta.

Pinevich, A. V. and Chunaev, A. S. Some problems concerning taxonomy and phylogeny of prochlorophytes, chlorophyll b-containing prokaryotes. (In Russian) Zhurn. obshchei Biologii 46:533–540; 1985.

The main problems of taxonomy and phylogeny of the so-called prochlorophytes (chlorophyll b-containing prokaryotes) are considered. The interest of this new group is determined by both its uncertain taxonomic position and phylogenetic relatedness to green algal and higher plant's chloroplasts. The position of prochlorophytes in the system depends on the status of cyanobacteria (blue-green algae) as well as on a criterion which allows to discriminate these two groups. The prochlorophytes are to be considered as a separate branch of oxygenic photobacteria.

Pugh, T. D. Studies in plant cell biology. I. Ultrastructure of algal and other prokaryotic associates of didemnid ascidians. II. Leaf ultrastructure of plants with crussulacean acid metabolism. Ph.D. Thesis, University of Wisconsin-Madison; 1976.

In the research reported herein, the algae and bacteria-like microorganisms associated with three species of ascidians in the family Didemnidae (Phylum Chordata: Subphylum Tunicata) from the Great Barrier Reef were investigated by light and electron microscopy in order to provide information which may prove useful in understanding energy pathways and nutrient cycles in marine environments. It was shown that there are several novel aspects to the ultrastructure of the algae and bacteria-like organisms inhabiting the ascidian colonies.

All of the algae associated with the didemnid ascidians *Didemnum ternatanum, Diplosoma virens,* and *Lissoclinum molle* were found to be blue-green algae. The characteristic appearance of these cells both light and electron microscopically is generally similar to that of other blue-green algae. Although the large, coccoid algae from the three ascidian species are generally similar in fine structural appearance, they possess distinctive features that allow them to be identified in terms of their host species. While it is certain that these algae are blue-greens, their classification to species remains in doubt. Even though according to one widely used system of classification all of the algae appear to be the single blue-green species, *Anacystis aeruginosa,* the consistent fine structural differences between them suggests that three species are represented. The association of these algae with didemnid ascidians is of special interest since it greatly extends the host range of blue-green algae to include primitive members of the Chordate, and thus represents an association with marine animals considerably higher in the evolutionary scale than hitherto recognized.

Schulz-Baldes, M. and R. A. Lewin. Fine structure of *Synechocystis didemni* (Cyanophyta: Chroococcales) Phycologia 15:1–6; 1976.

The green-coloured unicellular alga *Synechocystis didemni,* which has a pigment complement like that of a chlorophyte, is shown by transmission electron microscopy to have a typical prokaryotic construction. The nuclear body is not membrane-bound; there are no mitochondria; the thylakoids are distributed throughout the cell, though often more or less stacked; there are large polyhedral bodies with a crystalline fine structure; and the cell wall is four-layered. In these features *S. didemni* seems to be a typical cyanophyte.

Schuster, G., G. C. Owens, Y. Cohen, and I. Ohad. Light independent phosphorylation of the chlorophyll a,b-protein complex in thylakoids of the prokaryote *Prochloron. In:* Advances in Photosynthesis Research, Vol. III (C. Sybesma, ed.), Martinus Nijhoff / Dr W. Junk Publishers, The Hague/Boston/Lancaster; 1984: 4.283–4.286.

A membrane-bound kinase, which phosphorylates the chlorophyll *a,b*–protein complex in vitro and does not require activation by light, is reported in *Prochloron* cells isolated from *Diplosoma virens* at Eilat, Israel.—*eds.*

Schuster, G., G. C. Owens, Y. Cohen, and I. Ohad. Thylakoid polypeptide composition and light-independent phosphorylation of the chlorophyll *a,b*-protein in *Prochloron,* a prokaryote exhibiting oxygenic photosynthesis. Biochimica et Biophysica Acta, 767:596–605; 1984.

Thylakoids of the prokaryote *Prochloron,* present as a symbiont in ascidians isolated from the Red Sea at Eilat (Israel), showed polypeptide electrophoretic patterns comparable to those of thylakoids from eukaryotic oxygen-evolving organisms. Low temperature, fluorescence spectroscopy of *Prochloron*, having a chlorophyll *a/b* ratio of 3.8–5, and frozen in situ, demonstrated the presence of Photosystem II chlorophyll-protein complex emitting at 686 and 696 nm, as well as the emission band of Photosystem I at 720 nm which was so far not observed in *Prochloron* species. The latter emission was absent, if the cells or thylakoids were isolated prior to freezing. Energy transfer from chlorophyll *b* to chlorophyll *a* could be demonstrated to occur in vivo. The chlorophyll *a,b*-protein complex of Photosystem II, isolated by non-denaturing polyacrylamide gel electrophoresis, contained one major polypeptide of 34 kDa. The polypeptide was phosphorylated in vitro by a membrane-bound protein kinase which was not stimulated by light. A light-independent protein kinase activity was also found in isolated thylakoids of another prokaryote, the cyanophyte *Fremyella diplosiphon*. State I–State II transition could not be demonstrated in *Prochloron* by measurements of modulated fluorescence intensity in situ. We suggest that the presence of a light-independent thylakoid protein kinase of *Prochloron*, collected in the Red Sea at not less than 30 m depth, might be the result of an evolutionary process whereby this organism has adapted to an environment in which light, absorbed preferentially by Photosystem II, prevails.

Schuster, G., R. Nechushtai, N. Nelson, and I. Ohad. Purification and composition of photosystem I reaction center of *Prochloron* sp., an oxygen-evolving prokaryote containing chlorophyll *b*. F.E.B.S. Letters 191:29–33; 1985.

The photosystem I reaction center complex of the photosynthetic prokaryote *Prochloron*, present as a symbiont in ascidians from the Red Sea at Eilat (Israel), was isolated and characterized. The complex consists of 4 polypeptide subunits, and therefore is similar to that of cyanophytes and green algae. Their apparent molecular masses (in kDa) are about 70 (subunit I), 16 (subunit II), 10 (subunit III), and 8 (subunit IV). The purified reaction center contains about 40 chlorophyll molecules per P–700 as compared to about 800 in the intact thylakoids. Subunit I has the same apparent electrophoretic mobility and appears as a double polypeptide band as reported for this subunit in higher plants and algae. Immunological cross-reactivity was detected among subunits I and II of *Prochloron* and photosystem I reaction centers of higher plants and algae.

Seewaldt, E. and E. Stackebrandt. Partial sequence of 16S ribosomal RNA and the phylogeny of *Prochloron*. Nature 295:618–620; 1982.

Prochloron, a symbiont of colonial ascidians, is a unicellular prokaryote, containing chlorophylls *a* and *b* and lacking phycobiliproteins[1], thus differing from all other prokaryotes found, and resembling the chloroplasts of Euglenaphyceae, Chlorophyceae and those of higher plants in respect of their photosynthetic pigments. Some workers in the field of endosymbiosis already favour the idea that the green algal chloroplasts may have arisen by the uptake of prochlorons as symbionts[2–5]. Partial sequence analysis of 16S ribosomal RNA of *Prochloron* in fact assigns this organism to the cyanobacteria–chloroplast line of descent[6–8]. However, we report here that a comparison of 16S ribosomal RNA sequences detects no specific relationship between *Prochloron* and the chloroplasts of the green algae *Euglena gracilis* and *Chlamydommonas reinhardii*, nor between *Prochloron* and the chloroplasts of higher plants, represented by *Lemna* and *Zea mays*. On the contrary, *Prochloron* shows the highest sequence homology with *Nostoc, Fischerella, Agmenellum, Aphanocapsa* and *Synechococcus*. These results question the significance of *Prochloron* in the current view of the origin of chloroplasts.

Smith, H. G. On the presence of algae in certain Ascidiacea. Ann. Mag. Nat. Hist. (10 ser.) 15:615–626; 1935.

Out of ten species of compound Ascidians examined, only those from tropical waters (namely, *D. virens, D. voeltzkowi, T. cyclops,* and *D. viride*) were found to contain algae. In no case were algae found within the tissues of the animals. With the exception of *D. viride,* they were all situated in the cloacal cavities of the test. In *D. viride* the algae were embedded in the test, and were never found in the cloacal cavities. The significance and the origin of the relationship between the plant-cells and the Ascidians have been discussed.

Stackebrandt, E., and O. Kandler. The murein type of *Prochloron.* Zbl. Bakt. Hyg., I. Abt. Orig. 3 3:354–357; 1982.

Cell wall preparations were obtained from 4 batches of *Prochloron* isolated from 4 different species of ascidians. All preparations contained the typical murein components, glutamic acid, alanine, meso-diaminopimelic acid, glucosamine and muramic acid. Their molar ratios as well as the "fingerprints" of partial acid hydrolysates and the results of dinitrophenylation indicate strongly that the murein of *Prochloron* belongs to the variation A1γ (mDpm direct type) described by *Schleifer* and *Kandler* (1972).

Stackebrandt, E., E. Seewaldt, V. J. Fowler, and K-H Schleifer. The relatedness of *Prochloron* sp. isolated from different didemnid ascidian hosts. Arch. Microbiol. 132:216–217; 1982.

The relationship of *Prochloron* sp. isolated from four different didemnid ascidian hosts, namely *Lissoclinum patella, Lissoclinum voeltzkowi, Diplosoma virens* and *Trididemnum cyclops* was elucidated by comparative analysis of their 16S ribosomal ribonucleic acid (RNA). The oligonucleotide catalogues of the 16S rRNA obtained are almost identical, indicating a very close relationship among the prochlorophytes investigated. Phylogenetically *Prochloron* is a member of Cyanobacteriales.

Stam, W. T., S. A. Boele-Bos, and B. K. Stulp. Genotypic relationships between *Prochloron* samples from different localities and hosts as determined by DNA-DNA reassociations. Arch. Microbiol. 142:340–341; 1985.

Genotypic relationships between seven *Prochloron* samples isolated from different didemnid ascidian hosts collected at the Palau archipelago and Munda (Solomon Islands) and one cyanobacterial (*Synechocystis*) strain were determined by DNA-DNA reassociations. Thermal stability values of DNA-DNA hybrids indicate that all *Prochloron* samples involved are mutually very closely related and only slightly related with the *Synechocystis* strain. It is concluded that the *Prochloron* samples tested are representatives of one and the same species.

Summons, R. E. Occurrence, structure and synthesis of 3-(N-methylamino)glutaric acid, a new amino acid from *Prochloron didemni.* Phytochemistry 20:1125–1126; 1981.

3-(*N*-Methylamino)glutaric acid has been identified as a new free amino acid in extracts from *Prochloron didemnii* (Lewin), a unique prokaryotic algal symbiont associated with certain didemnid ascidians. Its structure was established by elucidation of the mass spectra of its TMSi and other derivatives and confirmed by synthesis.

Thinh, L.-V. Photosynthetic lamellae of *Prochloron* (Prochlorophyta) associated with the ascidian *Diplosoma virens* (Hartmeyer) in the vicinity of Townsville. Aust. J. Bot. 26:617–620; 1978.

Prochloron, the microscopic algae contained within the cloacal cavity of the ascidian *Diplosoma virens,* possesses fine structures similar to that of blue-green algae. However, the photosynthetic lamellae are composed of two appressed thylakoids, a feature slightly removed from the non-appressed thylakoids of prokaryotic blue-green algae.

Thinh, L.-V. *Prochloron* (Prochlorophyta) associated with the ascidian *Trididemnum cyclops* Michaelsen. Phycologia 18:77–82; 1979.

The unicellular algae harbouring within the cloacal cavities of *Trididemnum cyclops* (Michaelsen) possess both chlorophyll *a* and chlorophyll *b* (Chlorophyll *a*:*b* ratio of 10). They have a photosynthetic capacity of about 7 $\mu l\ O_2\ (\mu g\ chl\text{-}a)^{-1}\ h^{-1}$ and a light-saturated photosynthesis at a quantum flux density of 100–200 quanta $cm^{-2}\ sec^{-1}$.

The algal internal structure, resembling that of blue-green algae, consists of two definite zones bounded by a thin (30–50 nm), multilayered cell wall. The outer zone is occupied by the photosynthetic lamellae and the cytoplasm. The central zone is electron-transparent and sometimes contains lamellae of unknown nature. However, unlike single non-appressed thylakoids of the Cyanophyta, the algal photosynthetic lamellae are composed of two-appressed thylakoids.

The central zone undergoes binary division before cytokinesis. Cell division is by equatorial constriction.

Thinh, L.-V. and D. J. Griffiths. Studies of the relationship between the ascidian *Diplosoma virens* and its associated microscopic algae. I. Photosynthetic characteristics of the algae. Aust. J. Mar. Freshwater Res. 28:673–681; 1977.

The microscopic algae contained within the cloacal cavity of the ascidian *Diplosoma virens* have a high photosynthetic capacity [up to 5·3 μg C $(\mu g\ chl\ a)^{-1}\ h^{-1}$], giving the colony a high photosynthesis:respiration ratio (maximum 8·0–8·5). Photosynthesis is saturated at a quantum flux density of 120×10^{15} quanta $cm^{-2}\ s^{-1}$ (400–700 nm) and occurs at its maximum at temperatures between 30 and 35°C. The photosynthetic capacity is inhibited by chloramphenicol but not by cycloheximide. This is interpreted as indicating an affinity of these algae with the prokaryotes rather than the Chlorophyta, in spite of their possession of chlorophyll *b* (chlorophyll *a*:*b* ratio = 6·0) as an accessory pigment. In this and other respects the algae resemble those (*Prochloron*) described from other didemnid ascidians. Use of a buffered extraction medium has allowed separation from the animal tissue of photosynthetically active algal cells.

Thinh, L.-V., and D. J. Griffiths. *In vivo* absorption spectra for the prokaryotic algal symbionts in a range of didemnid ascidians of the Great Barrier Reef. Phycologia 22:93–95; 1983.

The *in vivo* spectral characteristics of *Prochloron,* isolated from five different ascidian symbionts, correspond more closely to those of green algae than to blue-green algae. In four of the five isolates there is an unexplained rising absorbance approaching the UV range of the spectrum below 360 nm.—*eds.*

Thinh, L.-V., D. J. Griffiths, and H. Winsor. Cellular inclusions of *Prochloron* (Prochlorophyta) associated with a range of didemnid Ascidians. Botanica Marina 28:167–177; 1985.

Cellular inclusions of sporadic occurrence are described for the *Prochloron* symbiont associated with a number of ascidian species belonging to the Didemnidae. Crystalline inclusions, showing a range of lattice periodicity (9 nm–29 nm), were of wide occurrence but were never seen in *Prochloron* from the two species of *Diplosoma* examined. Paracrystalline arrays of rods (11 nm in width, 7.2 nm apart) were restricted to *Prochloron* from three host species, *Trididemnum cyclops, Lissoclinum patella,* and *L. punctatum;* the latter host species also harboured *Prochloron* occasionally containing characteristic arrays of arcs or whorls. *Prochloron* from *Didemnum* aff. *candidum,* alone among those examined, contained distinctive bodies resembling starch grains whilst large, osmiophilic bodies were seen only in *Prochloron* from *Didemnum molle,* sometimes in great profusion.

Evidence is presented for the probable origin and pattern of growth of the crystalline and paracrystalline inclusions and it is concluded that these, and the other inclusions, play a role in the temporary storage of various cellular components. Because of their infrequent occurrence, the usefulness of these inclusions as taxonomic criteria is considered to be limited.

Thorne, S. W., E. H. Newcomb, and C. B. Osmond. Identification of chlorophyll *b* in extracts of prokaryotic algae by fluorescence spectroscopy. Proc. Natl. Acad. Sci. USA 74:575–578; 1977.

Solvent extracts of three different prokaryotic algae from three species of didemnid ascidians contained pigments identified, on the basis of their fluorescence excitation (E) and fluorescence emission (F) spectral maxima (measured in nm) at 77 K, as chlorophyll *a* (E 449, F 678) and chlorophyll *b* (E 478, F 658). The release of algae on cutting or freezing *Diplosoma virens* was accompanied by a strong unidentified acid that converted these pigments to pheophytins. This unexpected finding provided further confirmation of the identity of the chlorophylls on the basis of the fluorescence spectra at 77 K of pheophytin *a* (E 415, F 669) and pheophytin *b* (E 439, F 655). There was no evidence for the presence of the fluorescent bilin pigments found in other prokaryotic blue-green algae. Chlorophyll *a*/*b* ratios ranged from 2.6 to 12.0 in algae from different ascidians. The photosynthetic membranes were not organized into appressed thylakoids or grana in the algae from any of the three species of ascidians. The relationship between these observations and those in higher eukaryotic organisms is discussed.

Tseng, C. K., and B. C. Zhou. Some problems on the evolution of algae. Proceedings of the 1st Chinese Phycological Symposium, pp. 1–6; 1980.

The phylogeny and classification of the Plant Kingdom had been discussed in details in previous papers, in which we proposed to divide the Plant Kingdom into five subkingdoms, namely, the Bacteriophyta, Rhodocyanophyta, Chromophyta, Euchlorophyta and Mycophyta. Three subkingdoms, the Rhodocyanophyta, Chromophyta and Euchlorophyta are composed of what we call the algae, which are in turn divided into twelve phyla: Cyanophyta and Rhodophyta (Rhodocyanophyta), Cryptophyta, Dinophyta, Xanthophyta, Chrysophyta, Bacillariophyta and Phaeophyta (Chromophyta), Prochlorophyta, Euglenophyta, Chlorophyta and Charaphyta (Euchlorophyta).

It is believed that the structure and function of photosynthetic pigments and organelles are of primary importance in the early stages of the evolution of plants, and are therefore considered before the other non-photosynthetic characteristics in discussing phylogeny and devising systems of classification of the plants. In the present paper, some problems on the evolution of the algae are discussed.

1. According to the characteristics of the pigments and photosynthesis of the blue-green algae, these organisms are regarded as a group of algae rather than bacteria as suggested by some bacteriologists.

2. The eucaryotic algae are believed to have originated from common ancestors and then evolved along three different lines from three procaryotic ancestors, giving rise to the Rhodocyanophyta, Chromophyta and Euchlorophyta. The evolution took place in different evolutionary pathways and on different evolution levels.

3. Lewin's Prochlorophyta is accepted as a member of the Euchlorophyta. It was suggested that they were derived from the hypothetical protoflagellate and most closely related to the eucaryotic algae with chlorophylls *a* and *b*.

4. For the ancestor of Chromophyta, the significance of the discovery of dinoflagellates with phycobiliproteins was emphasized. The existence of procaryotic algae with chlorophylls *a*, *c* and phycobiliproteins may be expected.

Tseng, C. K., and B. C. Zhou. On *Prochloron* and its significance in algal phylogeny. Proceedings of the Joint China-U.S. Phycology Symposium (C. K. Tseng, ed.). Science Press, Beijing, China, pp. 29–38; 1983.

Prochloron occupies a very important position in the evolution of the plant world since it is the stock from which the chlorophytes originated. The authors discuss four stages whereby this may have occurred.—*eds.*

Van Valen, L. M. Phylogenies in molecular evolution: *Prochloron*. Nature 298:493–494; 1982.

Prochloron seems to have undergone relatively little evolution in its rRNA since it diverged from the ancestors of blue-green algae and green chloroplasts.—*eds.*

Whatley, J. M. The fine structure of *Prochloron*. New Phytol. 79:309–313; 1977.

The fine structure is described of *Prochloron*, a unicellular prokaryotic alga which contains both chlorophyll *a* and chlorophyll *b*. The cell wall resembles that of a blue-green alga. The thylakoids and cytoplasm together occupy a wide peripheral band. However, the thylakoids are present not as single lamellae, as in blue-green algae, but in pairs or, sometimes, in thicker stacks. Both thylakoids and cytoplasm are absent from a large central zone which is generally electron-transparent, but may contain electron-dense granules and fibrils. The function of the central zone is not known.

Withers, N. W., R. S. Alberte, R. A. Lewin, J. P. Thornber, G. Britton, and T. W. Goodwin. Photosynthetic unit size, carotenoids, and chlorophyll-protein composition of *Prochloron* sp., a prokaryotic green alga. Proc. Natl. Acad. Sci. USA 75:2301–2305; 1978.

Six samples of the prokaryotic, unicellular algae *Prochloron* sp., which occur in association with didemnid ascidians, were collected from various localities in the tropical Pacific Ocean, and their pigments and chlorophyll–protein complexes were identified and characterized. No phycobilin pigments were detected in any of the species. Chlorophylls *a* and *b* were present in ratios of $a/b = 4.4$–6.9. The major carotenoids were β-carotene (70%) and zeaxanthin (20%). Minor carotenoids of one isolate were identified as echinenone, cryptoxanthin, isocryptoxanthin, mutachrome, and trihydroxy-β-carotene; no ε-ring carotenoids were found in any sample. Except for the absence of glycosidic carotenoids, the overall pigment composition is typical of cyanobacteria. A chlorophyll *a*/*b*–protein complex was present in *Prochloron*; it was electrophoretically and spectrally indistinguishable from the light-harvesting chlorophyll *a/b*–protein of higher plants and green algae. It accounted for 26% (compared to ~50% in green plants) of the total chlorophyll; 17% was associated with a P700-chlorophyll *a*–protein. The photosynthetic unit size of 240 ± chlorophylls per P700 in *Prochloron* was about half that of eukaryotic green plants. A model is proposed for the *in vivo* organization of chlorophyll in *Prochloron*.

Withers, N. W., W. Vidaver, and R. A. Lewin. Pigment composition, photosynthesis and fine structure of a non-blue-green prokaryotic algal symbiont (*Prochloron* sp.) in a didemnid ascidian from Hawaiian waters. Phycologia 17:167–171; 1978.

Cells of a unicellular, green-coloured alga found growing endozoically in a didemnid ascidian around Hawaiian shores have been shown by electron microscopy to be clearly prokaryotic, like blue-green algae. The cell wall is like that of cyanophytes and there are no organelles. However, the thylakoids tend to occur in pairs or stacks, as in green eukaryotic algae. They contain no detectable amounts of bilin pigments (such as those characterizing blue-green algae) but have two distinct chlorophylls, chromatographically and spectroscopically identified as chlorophyll *a* and chlorophyll *b* (a combination characterizing green plants). Both intact didemnids, containing the algae cells and isolated algae cells were capable of vigorous photosynthetic oxygen production in light. These features suggest that the algal endosymbiont discussed here should be referred to the new algal division, Prochlorophyta.

Zeng, C. K., B. C. Zhou, A. S. Sun, Z. Z. Pan, and R. P. Zang. A preliminary report on *Prochloron* from China. Kexue Tongbao 27:778–781; 1982.

Prochloron is reported from several of the Xisha Islands and from Hainan Island, off the coast of China. Two chlorophylls, *a* and *b*, were identified by thin-layer chromatography: they occur in the ratio *a/b* = 5.6.—*eds.*

Other Published Articles on Didemnids Associated with *Prochloron* etc.

Cox, G. and Dibbayawan, T. 1987 A chloroplast-like DNA arrangement in *Synechocystis trididemni* (Chroococcales, Cyanophyta) Phycologia 26(1):148–151.

Eldredge, L. G. A taxonomic review of Indo-Pacific didemnid ascidians and descriptions of twenty-three central Pacific species. Micronesica 2:161–261; 1967.

Kott, P. Algal supporting didemnid ascidians of the Great Barrier Reef. Proceedings of the Third International Coral Reef Symposium, sponsored by the University of Miami, Rosenstiel School of Marine and Atmospheric Science; The Smithsonian Institution; the U.S. Geological Survey; May, 1977; Miami, Florida. pp. 616–622; 1977.

Kott, P. The ascidians of the reef flats of Fiji. Proc. Linn. Soc. N.S.W. 105:147–212; 1981.

Kott, P. Didemnid-algal symbiosis: algal transfer to a new host generation. Proc. Fourth International Coral Reef Symposium, Manila, 2:721–723; 1981.

Kott, P. Didemnid-algal symbioses: Host species in the western Pacific with notes on the symbiosis. Micronesica 18(1):95–127. 1982.

Kott, P., D. L. Parry, and G. C. Cox. Prokaryotic symbionts with a range of ascidian hosts. Bull. Mar. Sci. 34:308–312; 1984.

Kremer, B. P. *Prochloron*—neue Kategorie im System der Algen. Mikrokosmos 3:83–85; 1980.

Kruk, Jerzy. Prochloron a problem ewolucji chloroplastów. Wiadomości Botaniczne 31(1):29–42; 1987.

Lewin, R. A. La prochlorophyta; eble nove trovita subfilono de plantoj. Sciencaj Komunikaĵoj, 65–66, 1979.

Olson, R. R. Ascidian-*Prochloron* symbiosis: the role of larval photoadaptations in midday larval release and settlement. Biol. Bull. 165:221–239; 1983.

Parry, D. and Kott, P. 1988 Co-symbiosis in the Ascidiacea. Bull. Marine Science 42(1):149–153.

Rützler, K. 1981 An unusual bluegreen alga symbiotic with two new species of *Ulosa* (Porifera: Hymeniacidonidae) from Carrie Bow Cay, Belize. P.S.Z.N. I: Marine Ecology, 2(1):35–50.

Ryland, J. S., R. A. Wigley, and A. Muirhead. Ecology and colonial dynamics of some Pacific reef flat Didemnidae (Ascidiacea). Zool. J. Linnean Soc. 80:261–282; 1984.

Thinh, L.-V., D. J. Griffiths, and Y. Ngan. Studies of the relationship between the ascidian *Diplosoma virens* and its associated microscopic algae. II. Aspects of the ecology of the animal host. Aust. J. Mar. Freshwater Res. 32:795–804; 1981.

Tokioka, T. Ascidians found on the mangrove trees in Iwayama Bay, Palao. Palao Tropical Biological Station Studies, 2:499–506; 1942.

Appendix 1

List of *Prochloron* Expeditions

1. Gulf of California, Mexico, aboard RV "Dolphin": March, 1974: Meinhard Schulz-Baldes (W. Germany), Lanna Cheng (LC) and Ralph A. Lewin (RAL).
2. Hawaiian Institute of Marine Biology, Coconut Island, Hawaii: February 1976: William Vidaver (Canada), Nancy W. Withers (U.S.A.) and RAL.
3. University of Singapore: April, 1977: Linda Bonen (Canada), Kok-Kee Tan (Singapore), William Vidaver (Canada), Jean M. Whatley (England), LC and RAL.
4. Galápagos Islands, aboard RV "Alpha Helix": January–February, 1978: Andrew A. Benson (U.S.A.), Genèvieve Duclaux (France), Francoise Lafargue (France), David C. Smith (England), LC and RAL.
5. Western Caroline Islands, aboard RV "Alpha Helix" and at the Micronesia Mariculture Demonstration Centre (M.M.D.C.), Koror, Palau: September–October, 1979: Dilwyn J. Griffiths (Australia), Jimmy L. V. Thinh (Australia), R. L. (Ted) Pardy (U.S.A.), Ellen Weaver (U.S.A.), LC and RAL.
6. M. M. D. C., Palau: June, 1981: Charles Birkeland (Guam), R. L. (Ted) Pardy (U.S.A.), Don Phipps (U.S.A.), Erko Stackebrandt (W. Germany), David S. Woodruff (U.S.A.), LC and RAL.
7. M. M. D. C., Palau: February–March, 1982: Randall S. Alberte (U.S.A.), Lana Fall (U.S.A.), Ray Fall (U.S.A.), Grover C. Stephens (U.S.A.), LC and RAL.

8. M. M. D. C., Palau: February–March, 1983: Randall S. Alberte (U.S.A.), Jaime Matta (Puerto Rico), Hans W. Paerl (U.S.A.), Mitsu Okada (Japan), Hewson Swift (U.S.A.), LC and RAL.
9. M. M. D. C., Palau, March, 1988: Randall S. Alberte (U.S.A.), Hewson Swift (U.S.A.), Richard C. Zimmerman (U.S.A.), LC and RAL.

Appendix 2

First *Prochloron* Workshop

The First International *Prochloron* workshop* was held at the Scripps Institution of Oceanography in La Jolla, California, January 3–6, 1983. There were some 42 participants, including five from Australia; four from Canada; one each from West Germany, Israel, The Netherlands, and the United Kingdom; and 29 Americans (11 local, the rest from other parts of the U.S.A.). Topics discussed included ecology, physiology, and nutrition, fine structure, pigments and other lipids, enzymes, nucleic acids, and phylogeny. A short history of "prochlorophycology" was presented, and plans for future collaborative research on *Prochloron* were discussed. The following is a list of invited participants.

Randall S. Alberte, USA
Naval Antia, Canada
William R. Barclay, USA
Andrew A. Benson, USA
David G. Bishop, Australia
Douglas Bruce, Canada
David J. Chapman, USA
Lanna Cheng, USA
Yehuda Cohen, Israel
Guy Cox, Australia
Elisa d'Amelio, USA
Teresa Dibbayawan, Australia
Robert C. Fahey, USA
Ray Fall, USA
Tom H. Giddings, USA
Francis T. Haxo, USA
Robert W. Hoshaw, USA
Barbara Javor, USA
Janette Kenrick, Australia
Ralph A. Lewin, USA
Lynn Margulis, USA
Richard M. McCourt, USA
Bruce A. McFadden, USA
James Marshall, USA

* Funded by the National Aeronautics and Space Administration, U.S.A.

Tony F. Michaels, USA
Alan Mileham, USA
Gerald L. Newton, USA
Hans W. Paerl, USA
R. L. (Ted) Pardy, USA
Gregory M. Patterson, USA
Radovan Popovic, Canada
Michael R. Silverman, USA
Douglas W. Smith, USA
Erko Stackebrandt, West Germany
Wytze T. Stam, The Netherlands
Hewson Swift, USA
L. V. (Jimmy) Thinh, Australia
Robert K. Trench, USA
William Vidaver, Canada
Ellen C. Weaver, USA
P. David Weitzman, England
David S. Woodruff, USA

Author Index

Bibliographic citations in italics

Subject Index

* According to international codes of nomenclature, the specific epithets should be spelled *"voeltzkowii"* and *"lampertii"*. *Eds.*